K. Morioka

Hair Follicle

Differentiation Under the Electron Microscope

An Atlas

K. Morioka

Hair Follicle

Differentiation Under the Electron Microscope An Atlas

With 83 Figures, Including 6 in Color

Springer

Kiyokazu Morioka, Ph.D.
The Tokyo Metropolitan Institute of Medical Science
3-18-22 Honkomagome, Bunkyo-ku, Tokyo 113-8613, Japan

Library of Congress Control Number: 2004 112646

ISBN 4-431-22429-7 Springer-Verlag Tokyo Berlin Heidelberg New York

Printed on acid-free paper

Springer is a part of Springer Science+Business Media
springeronline.com

Typesetting: SNP Best-set Typesetter Ltd., Hong Kong
Printing and binding: Nikkei Printing Inc., Japan

Foreword

Each and every hair is much more than just the visible shaft—there are also associated complex sheath structures of epidermal and dermal origin. In the hair follicle, cells undergo a variety of differentiation processes, mostly depending on their layers and positions therein, and electron microscopy reveals a very complex architecture. The structure of a particular layer, such as Henle's layer of the inner root sheath, is not uniform. Rather, cells drastically change during the course of differentiation. By simply comparing electron micrographs of cells of a layer at different degrees of differentiation, one can hardly recognize them as belonging to the same layer.

As readers will see, this book contains many superb electron micrographs, from low-magnification panoramic views for orientation to high-power views showing ultrastructural detail. Captions and schematic drawings are also very helpful in "reading" electron micrographs and understanding the structural detail. In this way, Dr. Morioka has succeeded in dissecting the complex hair follicle at the ultrastructural level.

Dr. Morioka started his career as a biologist when he was a graduate student in the Laboratory of Developmental Physiology and Biochemistry at the Zoological Institute of the University of Tokyo. He continued to work in the field of biochemistry of cellular differentiation of blood cells before moving on to hair follicles. About 10 years ago he visited our laboratory to take a course in electron microscopy. I am sure that he learned not only the techniques of microscopy, but also the structural marvel of cells and tissues of the body.

After the completion of genome projects at the beginning of this century, complex tissue systems will provide a new challenge to biomedical researchers. Structurally, while hair is compact, it is one of the most complex cellular systems in the body. Hair serves as one of the outstanding models for studying cell renewal, proliferation, differentiation, migration, and death. Comprehensive understanding of the structure of hair constitutes the basis for the advancement of research in this field. Dr. Morioka's work on the hair follicle sets the stage for even deeper ultrastructural research into hair.

Advances in electron microscope technology and the use of diamond knives have made thin-section electron microscopy much easier in most specimens. However, some tissues are still difficult to process. Making smooth ultrathin sections from specimens containing different textures and degrees of hardness is not easy. Skin is one such composite tissue. High-quality electron micrographs can be obtained by highly skilled electron microscopists only by sectioning and processing with utmost care. It is felt that Dr. Morioka's achievement offers an admirable and comprehensive reference work and will benefit many researchers working in this field.

Kuniaki Takata, Ph.D., D.M.Sc.
Professor and Head
Department of Anatomy and Cell Biology
Gunma University Graduate School of Medicine
Gunma
Japan

Preface

New approaches to the study of hair, involving growth factors, differentiation inducers, cell signaling pathways, and transcription factors, have vastly increased our understanding of hair growth in recent years. Each hair contains stem cells and the hair's distinctive progenitors, primarily the medulla, cortex, cuticle, and three inner root sheath layers. The differentiation of these components is controlled by complicated systems, including epidermis–dermis interactions, which are known to function in a variety of developmental processes in humans and animals. It suggests that advances in hair research contribute not only to the artificial control of hair growth, but also to the elucidation of core problems in biology and medicine.

About 10 years ago, my friend Prof. S. Ihara (Faculty of Life and Environmental Science, Shimane University), who had studied hair reproduction in vitro, introduced me into his mysterious and fascinating field. His work and thinking facilitated my early studies, but I soon noticed that the structure of hair is highly complex and, therefore, sufficiently detailed descriptions of the morphological changes in each hair-cell lineage and the interactions of cells of different lineages were not available. Many questions emerged. How and when do uniform placode cells differentiate? What conditions are needed to induce the intrusion of dermal papilla cells into the hair rod? How do connective tissue follicle (dermal sheath) cells transform into the papilla? How does the hair bulb stem cell differentiate into several lines of very different types of cells? How do medulla cells differentiate and undergo programmed cell death (and is this a common apoptotic event)? How are the hard keratins assembled in the hair cortex and hair cuticle? How are the distinctive cornifications of the three inner root sheath layers accomplished? Many more questions appear in the text of this book.

I thought that some of these questions might be answered, at least partially, by recourse to electron microscopy. At that time I determined to begin the study of ultrastructure. I was not young, but Prof. M. Sameshima (Faculty of Agriculture and Life Science, Hirosaki University) took me under his wing as a rank beginner and tutored me broadly on the methods of electron microscopy, from the fixation of samples to the printing of photographs. Prof. K. Takata (Gunma University, Graduate School of Medicine) also gave me many useful suggestions on his professional training course in this field. With their lessons absorbed, I entered into an intensive investigation of hair.

Because it is refractory to the penetration of resins, hair is one of the most difficult organs to deal with in electron microscopy. I decided to use neonatal rats, partly because their relatively soft hair is easier to work with, but more importantly, because the use of animals at this stage of life allows us to see the developmental process of the hair follicle. Hair shafts regenerate throughout the life of humans and animals, but hair follicles are not formed in adults. The hair follicle of a rat from birth to seven days does not contain two (old and new) hair shafts because it is in the first growth stage, the first anagen. The advantage of this characteristic is that it obviates the need to identify old and new hair shafts in the small viewing field of the electron microscope.

Electron microscopy research of hair has a history of some 50 years, but it has become difficult to gather and integrate the fragmented knowledge so far accumulated. Some has been lost or become inaccessible during the age of molecular biology. So, I chose to provide various new and original photographs of all basic hair cells, from their birth to death.

As for the intended audience for this book, first and foremost, I recommend this collection to molecular biologists who are interested in hair. A hair is a minute organ, but it has a fine structure consisting of many distinctive tissues and cells and is produced through the collaboration of these constituent members. All those who study hair should know the morphological and developmental details of the organ. Some photographs that contain novel findings may also be useful to advanced students and experts conducting morphological or dermatological research. Chapter 1 will be of benefit to students or novices in this field. It provides a brief history of hair and hair research as well as a short summary of recent progress in the field. It also contains basic information and nomenclature to facilitate the beginner's study. The following chapters contain brief explanations of the photographs, but it is important to emphasize that the photographs likely contain more information than I have described. I therefore expect that the readers will discover new possibilities and questions in the photographs themselves, aside from any explanations I have offered.

I express my sincere thanks and appreciation to the following people for their support and interest during the course of this work: Prof. M. Sameshima and Prof. K. Takata for teaching me the principles and techniques of electron microscopy and for writing the foreword to this book; Prof. S. Ihara and Dr. T. Matsuzaki for their many helpful suggestions; Dr. I. Yahara for leaving his electron microscope for us when he retired; Dr. H. Sakuraba, the director of our laboratory, for allowing me to undertake this research; Dr. H. Takano-Ohmuro, Dr. M. Mukaida, Dr. E. Kominami, Dr. T. Ueno, Dr. M. Sameshima, Dr. K. Sato-Kusubata, Dr. S. Kawashima, Dr. H. Sakuraba, and Dr. S. Ihara for their collaboration on the immunohistochemical study shown in Chapter 1; Mr. M. Kawasaki for his goodwill in allowing the use of his original drawings of ancient animals; Dr. S. Kimura-Nozawa for her generous gift of the book *Hair*, written by M. L. Ryder and translated into Japanese by Dr. Kimura-Nozawa with the late Dr. T. Kato; and Mrs. S. Yasutomi-Sakaizumi for the schematic of the hair follicle structure and Dr. G. Harris for his rewriting and editing of the text, both of whom worked above and beyond the call of duty. I also thank the editorial and production staff of Springer-Verlag Tokyo, who provided exceptional help in the publication of this book. To all of you, I gratefully acknowledge your efforts.

Finally, I express special thanks to Dr. M. Morioka for her encouragement throughout this study and my gratitude also goes to my experimental animals, which died in their infancy. I dedicate this book to them.

Kiyokazu Morioka
Shiroi, Chiba
Japan

Contents

3

Hair Cortex and Hair Cuticle 45

4

Inner Root Sheath 63

5

Outer Root Sheath and Companion Layer 89

Introduction 1

Why Study Hair?

During the 10-minute walk from our institute to the local railway station, Tabata, the acute observer will identify at least nine barbershops and hair salons on the way. It is said that Japan boasts more than 350,000 shops in the business of hair. And although one may think there is a convenience store on every street corner, their total number is only about 40,000. No matter with what little regard they might hold their other tissues and organs, there can be scarcely a single human (or beast) alive who does not attend to their hair at least once a day.

From an evolutionary point of view, the acquisition of hair may be second only to the development of the amnion as the most remarkable event in the history of animals. It has enabled them to remain active in the cold night air and to thrive in the Earth's higher latitudes. Even the once-dominant dinosaur was forced to evolve toward the feathered bird (although the origins of the feather and mammalian hair are thought to be different) [1]. The history of mammals witnesses the primacy of hair, its greater importance and utility, over any weapon, armor, or massive body size, because now all but a few (such as pangolins) boast a coating of the stuff.

Humans are a special species with regard to hair. It seems we have reduced its value to the level of insignificant accessory. It may yet play a modest physiological role in protecting the head from concussion and ultraviolet irradiation and in providing warmth and discharging toxic substances. No matter; it is hairstyling, canities (graying), hypotrichosis (thinning), and alopecia (hair loss) that have engaged the cosmetic, medical, and academic interests of investigators, and the interest of many ordinary people too.

It is also notable that humans have many kinds of hair, from scalp hair, eyebrows, eyelashes, whiskers, and beard, to axillary and pubic hair, and pelage equivalents and lanugos. It remains to be determined how they are regulated under different control systems. Although the hair of humans may degenerate in the evolutionary timescale, the importance of research into the development of hair will not greatly diminish, because of its unique modeling value in studies of the processes of tissue development, cell differentiation, stem cell management, and programmed cell death, as are discussed in later sections.

Table 1.1. Fossil chronology

Millions of years ago (from)	Era	Period	Epoch	Animals (selected arbitrarily)
1.65	Cenozoic	Quaternary		Human
		Tertiary	Pliocene	
			Miocene	
			Oligocene	
			Eocene	
6.5			Paleocene	Mammals
				Birds
145	Mesozoic	Cretaceous		Bony fish
208		Jurassic		Dinosaurs
245		Triassic		
290	Paleozoic	Permian		Mammal-like reptiles
354		Carboniferous		Amphibia
408		Devonian		Jawed fish

Apart from any consideration of today's interests, hair and its relevant genes have a long paleontological history. The acquisition of hair by some Paleozoic reptiles is speculated to have dramatically changed their shape and lifestyle. The chronology of fossils is shown in Table 1.1. The evolution of hair may have been accelerated by the cold climate of the Permian period. Hair would have helped to maintain a constant body temperature and thus ensure the life of endothermic animals, which were small, agile, and nocturnal. By virtue of their hair, they were allowed to live and evolve in secret through the age of the dinosaurs, and even survive their extinction.

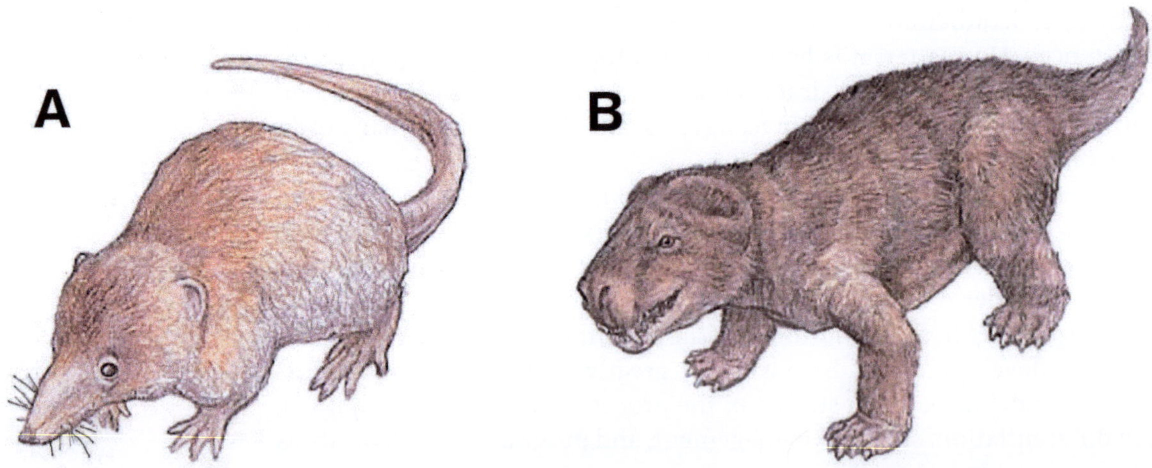

Fig. 1.1.

The earliest reliable record of hair is found in a fossil of the Paleocene period [2], in which the structure of hair cuticles is preserved. Its appearance suggests that the complicated structure of the hair follicle, closely similar to that of present-day mammals, had already appeared at this time. From a second point of view, the finding also highlights the long evolutionary history required to make this type of hair structure. Some paleontologists infer the earlier appearance of hair based on fossils of the megastrozodon (Fig. 1.1A) and other archaic mammals from the Late Triassic or Early Jurassic period in the Mesozoic era [3], although no direct evidence has been found.

If the possession of hair is a general characteristic of mammals including the megastrozodon, the advent of hair occurred around 200 million years ago at the latest. Furthermore, mammal-like reptiles such as the Mesozoic *Cynognathus* as well as the Paleozoic *Gorgonops* (Fig. 1.1B,C) and others might also have had hair [4]. This notion is based on their bone structure: the hollows around the nose of these species suggest the presence of sensory hairs [5, 6]. Here, I would like to quote a passage by Knafelc [5]: "Small holes (foramina) were beginning to be found on the skulls of some individuals. Nerves and blood vessels were believed to have passed through the tiny foraminas that led to active cheek and lip muscles, and even to whiskers. Although the fossil record does not display the origin of hair, many paleontologists believe that the gradual spread of sensitive whiskers may have been the start of hair." The opinions may be divided on the matter of the last phrase. I speculate that the pelage developed before the establishment of sensitive whiskers, because the need for pelage seems to be more fundamental, and because the structure of the sensory hair is more complex. It is also notable that the degeneration of the ribs of the lower torso enabled animals to groom themselves, in exchange for the risk of visceral wounds by carnivorous animals or accidents. If any Paleozoic mammal-like reptiles already had hair, it is not difficult to believe that some may have survived the coldness of the Permian extinction with the aid of their hirsute pelages. The survived special reptiles are thought to be the direct ancestor of mammals. Thus, the paleontological record suggests that hair has been essential to the survival of endothermic animals on earth.

C

Fig. 1.1. Illustrations of ancient animals believed to have had hairs. **A** Megastrozodon, one of the oldest mammals, appeared in the Triassic and Jurassic periods of the Mesozoic era. **B** *Cynognathus*, a mammal-like reptile that appeared in the Triassic period of the Mesozoic era, is deduced to be closely related to mammals. **C** *Gorgonops*, a mammal-like reptile, appeared in the Permian period of the Paleozoic era. (Illustrations used courtesy of Satoshi Kawasaki)

Growth and Development

For mammalian babies, hair is beneficial as insulation from the very moment of birth. The development of fetal hair therefore usually begins before birth. The fetuses of rats, my experimental animal in this work, start to grow hair buds in the dorsal skin around day 16 of gestation (Fig. 1.2B). The periderm that then covers the embryo is removed, and concurrently a cornified epidermis starts to develop. The epidermis develops as an organ with fully matured structures before birth (Fig. 1.2C). Some hair follicles with accompanying developed dermal papilla are also formed before birth (Fig. 1.2C). As shown in Fig. 1.2D,E, the number and length of hairs increase during the first week after birth. The

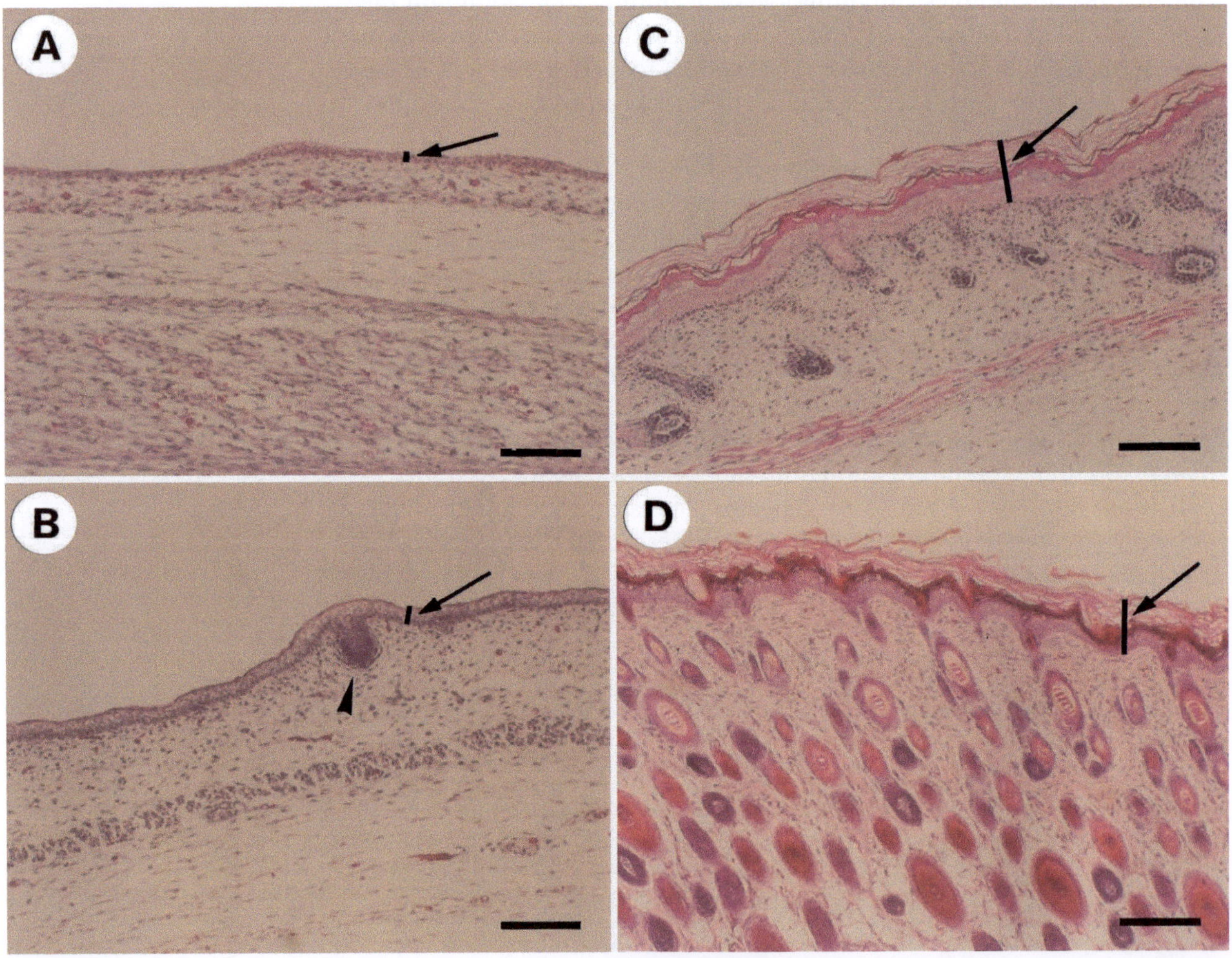

Fig. 1.2.

growth period of hairs (anagen) continues until the first regression (catagen), which occurs a few weeks later. Catagen means a transient degeneration of the hair follicle that involves the loss of the old hair shaft. The new hair shaft is regenerated from the same position of the renewed hair follicle after a short rest period (telogen) [7–9]. Most animals (including humans) show a daily cycle of hair loss and regeneration throughout life. If this mechanism was not available, every animal would drag a train of hairs behind it wherever it went. Detailed analysis of the hair cycle is beyond the scope of this book; references 8 and 10 will provide the interested reader with a closer understanding of the subject.

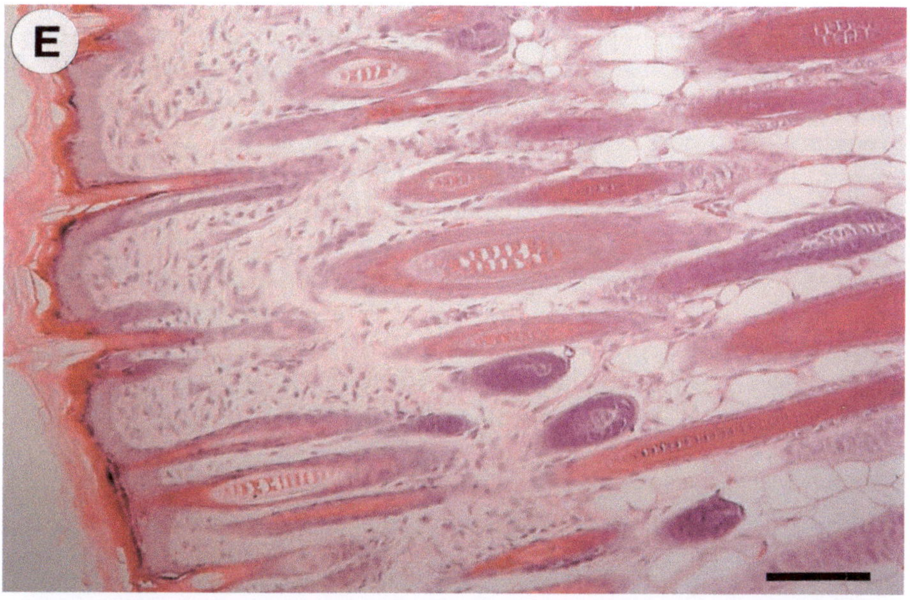

Fig. 1.2. Histological features of the skin of fetal rats on day 16 of gestation (**A**) and on day 18 of gestation (**B**), and those of neonatal rats on the day of birth (**C**) and on day 7 after birth (**D, E**). The day of birth is referred to as day 0. The photograph in **E** was taken with a digital camera system; reversal films (Provia 100F; Fujifilm) were used for the other photographs. The *arrowhead* in (**B**) represents a primitive hair follicle, which may be called a hair bud. *Vertical bars* pointed to by *arrows* represent the thickness of the epidermis, which may include the periderm in fetal animals. *Bars* 50 μm

The process of the first wave of hair growth in fetal and neonatal animals is schematically summarized in Fig. 1.3 based on the traditional classification of Hardy [11]. The downward invagination of the epidermis into the dermis is a special feature of hair formation and is distinct from the morphogenesis of feathers, in which the epidermis grows upward with no such invagination [1]. A number of investigators have speculated on the evolutionary relationships of these epidermal appendages [1, 12, 13], although they are not discussed here.

Recently, the induction of the hair bud has been shown to depend on the Wnt signaling pathway. "Wnt" proteins are thought to be involved in many differentiation events during animal development [14]. They are the extracellular regulatory ligand proteins that bind to the cell-surface receptor "Frz." Binding of Wnt proteins to the receptor activates some factors such as dishevelled that block the activity of the β-catenin–phosphorylation complex. The suppression of this complex reduces the phosphorylation of β-catenin, which leads to the stabilization of β-catenin [15–18]. Because β-catenin activates a transcription factor "Lef-1" to bind to the promoters, which is an essential process for hair formation, the stabilization of β-catenin is important for the production of the hair [19, 20]. It is intriguing that transient (not constitutive) suppression of the β-catenin destruction system, which is normally brought about by Wnt signaling, triggers the active growth phase of the hair cycle in mice [21, 66].

The synthesis of Lef-1 is dependent on noggin/BMP. The suppression of BMP (bone morphogenetic protein) activity by noggin induces the synthesis of Lef-1. The Lef-1/β-catenin complex suppresses the expression of E-cadherin, which promotes the conversion of cadherins from type E (epidermal) to type P (hair). Thus, noggin/BMP is thought to be important to the differentiation of hair progenitors from the skin cells [20, 22, 23]. Recent research revealed that the disruption of a BMP receptor causes impairment of the differentiation of inner root sheath [67, Fig. 1.4]. Of course, hair formation is suggested to depend not only on the Wnt/β-catenin and noggin/BMP systems but also on other factors [24–28]. Induction of the hair bud is one of the most active fields in current hair research; see Fuchs and colleagues for reviews [29, 30].

Fig. 1.3. Diagram of hair growth. Stages are described according to the definition of Hardy [11]. Areas of *horizontal stripes* represent the inner root sheath; *black areas* represent the hair shaft

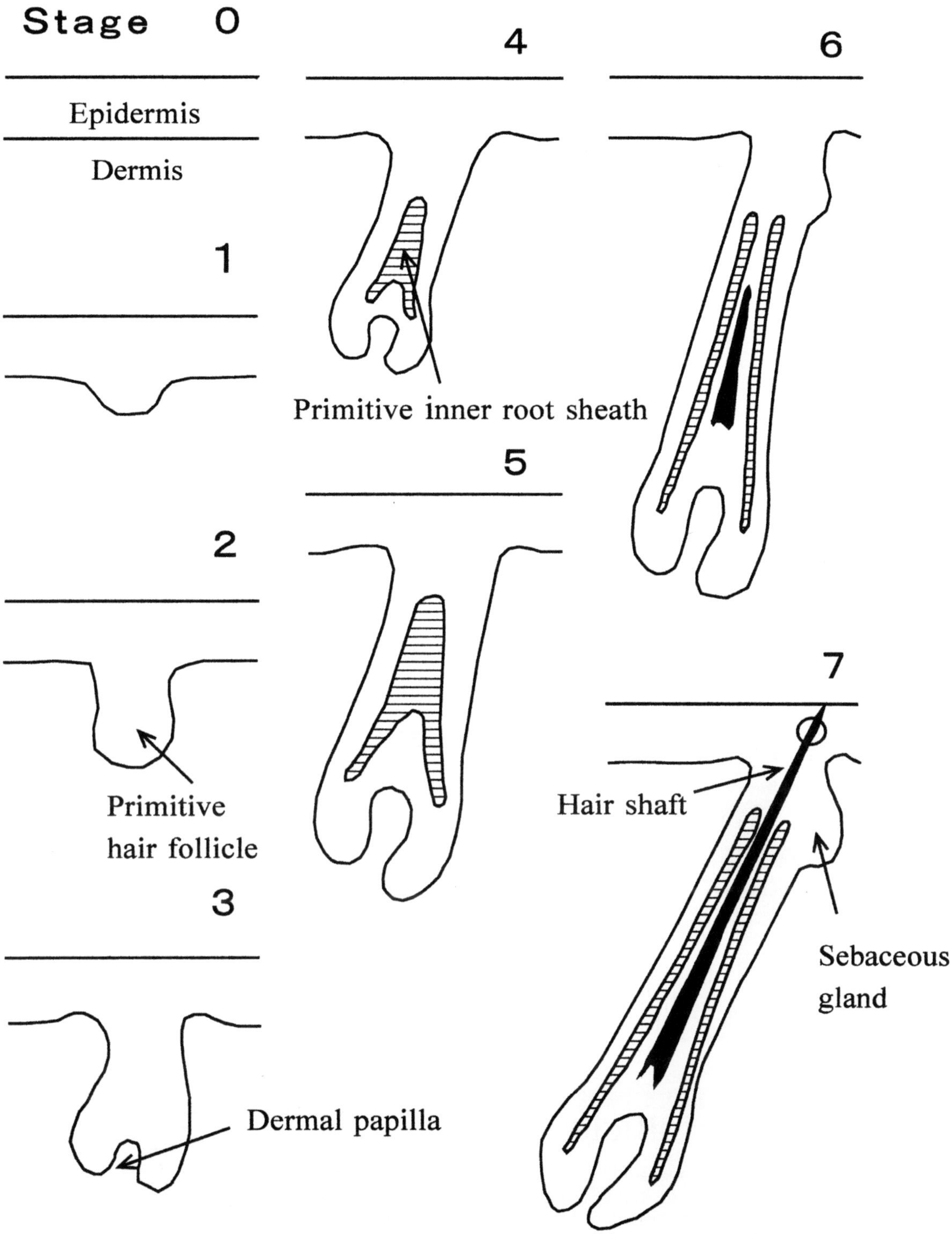
Stage 0
Epidermis
Dermis
1
2
Primitive hair follicle
3
Dermal papilla
4
Primitive inner root sheath
5
6
7
Hair shaft
Sebaceous gland

Fig. 1.3.

The growing hair bud is intruded by dermal cells from its lower (distal from the skin surface) end to form the dermal papilla (stage 3 in Fig. 1.3). It is notable that the areas of the inner root sheath shown in stages 4 and 5 do not include the sheath's whole lineage, as the inner root sheath is to be formed by cells that are derived from the hair bulb area, which is outside the area designated by horizontal stripes. The hair shaft is formed in stage 6 and extends to penetrate the skin epidermis in stage 7. The lower (approximately one-third) part of the hair in stage 6 is enlarged in Fig. 1.4 to show the details of structures, nomenclatures, and abbreviations. The abbreviations used in this book are based on those in Fig. 1.4 although I often use the unabbreviated terms to facilitate the understanding of readers new to the field.

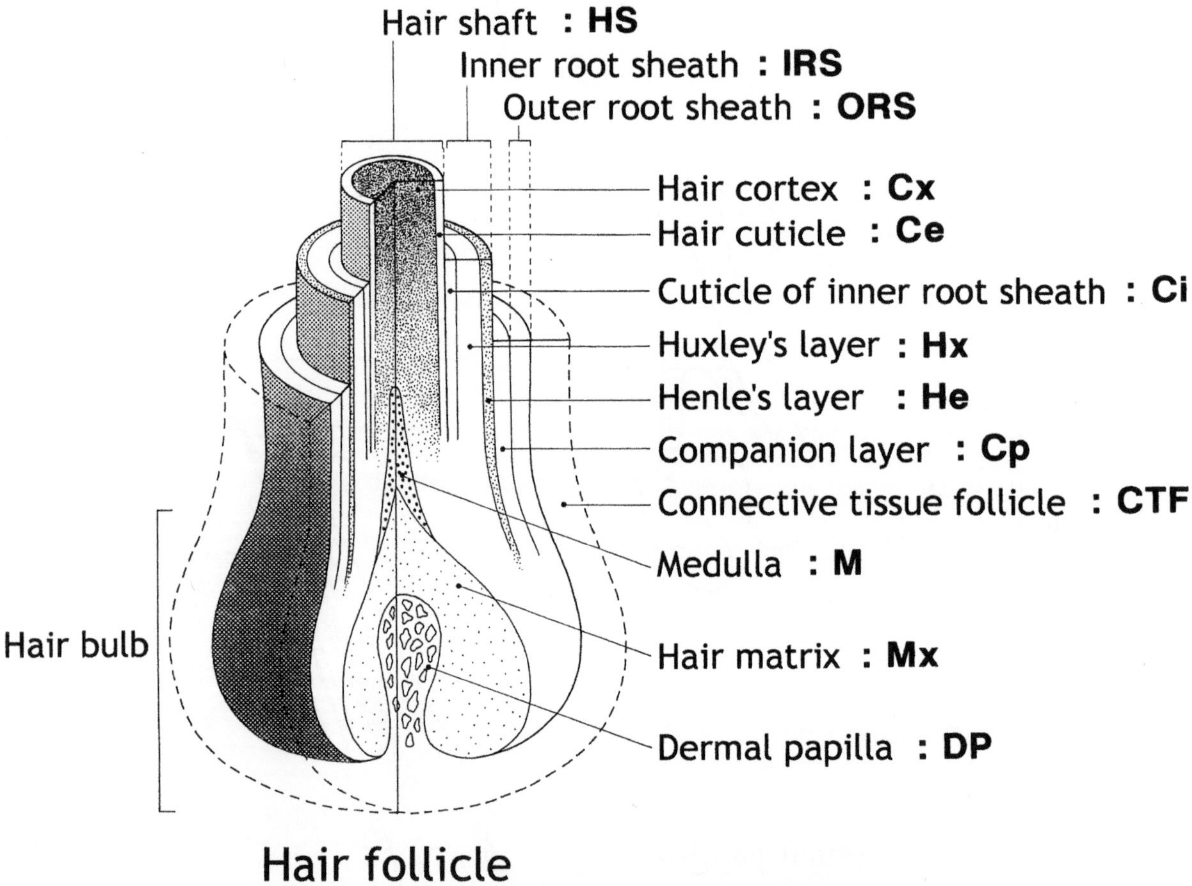

Fig. 1.4. Schematic representation and terminology of the hair follicle. Most of the abbreviations used in this book are shown in this figure. This part of the hair corresponds largely to the one-third of the hair follicle at stage 6–7 in Fig. 1.3. The hair shaft (*HS*) consists of medulla (*M*), hair cortex (*Cx*), and hair cuticle (*Ce*). The inner root sheath (*IRS*) consists of cuticle of inner root sheath (*Ci*), Huxley's layer (*Hx*), and Henle's layer (*He*). Although the companion layer (*Cp*) is occasionally referred to as a member of the outer root sheath (*ORS*), I use the terms companion layer and outer root sheath independently. I use *CTS* (connective tissue follicle) as having the same meaning with the dermal sheath. *Mx*, hair matrix; *DP*, dermal papilla

General Structure

The structure illustrated in Fig. 1.4 is believed to be basically common to most mammalian hairs, including lanugos and sensory hairs. The hair follicle consists of nine distinctive epidermal layers: Mx (hair matrix), M (medulla), Cx (hair cortex), Ce (hair cuticle), Ci (cuticle of the inner root sheath), Hx (Huxley's layer), He (Henle's layer), Cp (companion layer), and ORS (outer root sheath), arranged concentrically from core to periphery, as well as two dermal tissues: DP (dermal papilla) and CTF (connective tissue follicle; this term is used as the equivalent of the term dermal sheath in this book) [31–33]. Among these tissues, only the medulla is optional, given that some hairs have no medulla whereas in others it is relatively large.

The hair cortex (Cx) and hair cuticle (Ce) constitute the major part of the hair shaft that penetrates the skin. Both Cx and Ce tissues undergo heavy keratinization to form the solid hair shaft, although their components or protein species are different (Table 1.2). The three concentric layers located immediately distal (external) to the shaft constitute the inner root sheath (IRS), which is thought to support the growth and differentiation of the shaft. The innermost layer of the IRS, called the cuticle of the inner root sheath (Ci), consists of thin overlapping scales facing the hair

Table 1.2. Keratins in hair follicles

	Type I	Type II
Medulla	K 17, K 19 [34, 35]	K 5, K6hf and/or K 6 [34, 47]
Cortex	Ha 1, Ha 3, Ha 4 Ha 5, Ha 6, Ha 8 [36]	Hb 1, Hb 3, Hb 6 [37]
Cuticle	K 10, K 11 [38, 39] Ha 2, Ha 5 [36]	K 1, K 7 [38, 39] Hb 2, Hb 5 [37]
Inner root sheath	K 10 [39, 40], K 13 [35], K 18 [41] IRSa [42]	K 1 [39, 40], K 4 [41] K 6irs [41, 43]
Companion layer	K 14 [35], K 16, K 17 [40, 44] K 18 [41]	K 6 [40, 44], K 8 [35] K 6hf [45]
Outer root sheath	K 14 [35, 40, 41], K 15 [46], K 17 [41] K 19 [41]	K 5 [40, 41], K 6 [41]
Hair matrix	K 14 [35], K 15 [46], K 17 [34] Ha 5 [37]	Hb 5 [37], K6hf [47]

irs, inner root sheath; hf, hair follicle

cuticle. The conjunction of these two cuticles is special and is described in detail later. The intermediate Huxley's layer (Hx) is the last of the keratinizing tissues in the IRS and hair shaft to undergo this process. In other words, metabolically active cells in the Hx survive until the last and may possibly play a role in the care of other keratinized cells in terms of nutritional and informational support. At the least they help in maintaining plasticity to allay the distortion brought about by unbalanced keratinization of the shaft, such as occurs in curling. Further, by coordination with the cuticle of the inner root sheath, they also help maintain the smooth cylindricality of the hair follicle structure. When the IRS is absent, the hair cuticle may become entangled in the dermis or epidermis owing to its rough surface, as occurs with hangnails (Fig. 1.5; also see Fig. 4.16). Of course, the significance of Hx and of the other two layers of the IRS remains to be elucidated.

In sharp contrast to the Hx, Henle's layer (He) keratinizes at a very early phase of hair growth, and consequently keratinized He cells are observable at a positionally low level of the hair follicle. It seems that the inner cells, namely the rest of the IRS cells and all the hair shaft cells, grow and differentiate under the guardianship of the surrounding rampart of He. At the least this support provides mechanical strength to the delicate part of the follicle in the early stages. I speculate that unknown informational substances may be concentrated within the He rampart by some perturbation of diffusion or transport, which would at the same time evoke a corresponding risk of nutritional insufficiency in the inner cells. Essentially no angiogenesis occurs in the dermal papilla of the early follicle.

All layers of the IRS form trichohyalin granules and then keratinize, but they do not form a compact and enduring structure like the shaft. The cornified IRS layers finally desquamate together with the skin surface cells. The companion layer (Cp) [48] and outer root sheath (ORS) are the outermost cellular group of the follicle. The ORS is continuous with the basal layer of the epidermal epithelium, but it is incapable of producing spinous, granular, or cornified cells such as those found in the epidermis. Generally speaking, the Cp and ORS remain enigmatic with regard to both nature and role.

Fig. 1.5. An electron microscopic view of a hair follicle at a position slightly higher than the upper end of the hair bulb. The section was cut longitudinally. The hair grows in the *upward* direction. The sample was prepared from a day 0 neonatal rat. With the exception of some of the Henle's layer cells (*asterisks*) found around the *upper end* of the figure, cells are not keratinized at this height of the follicle. The *horizontal dual-arrow bar* (*HS*) indicates the area of the presumptive hair shaft. Boundaries of tissues are shown by *dotted lines* in the middle area of the figure. *Bar* 2 μm

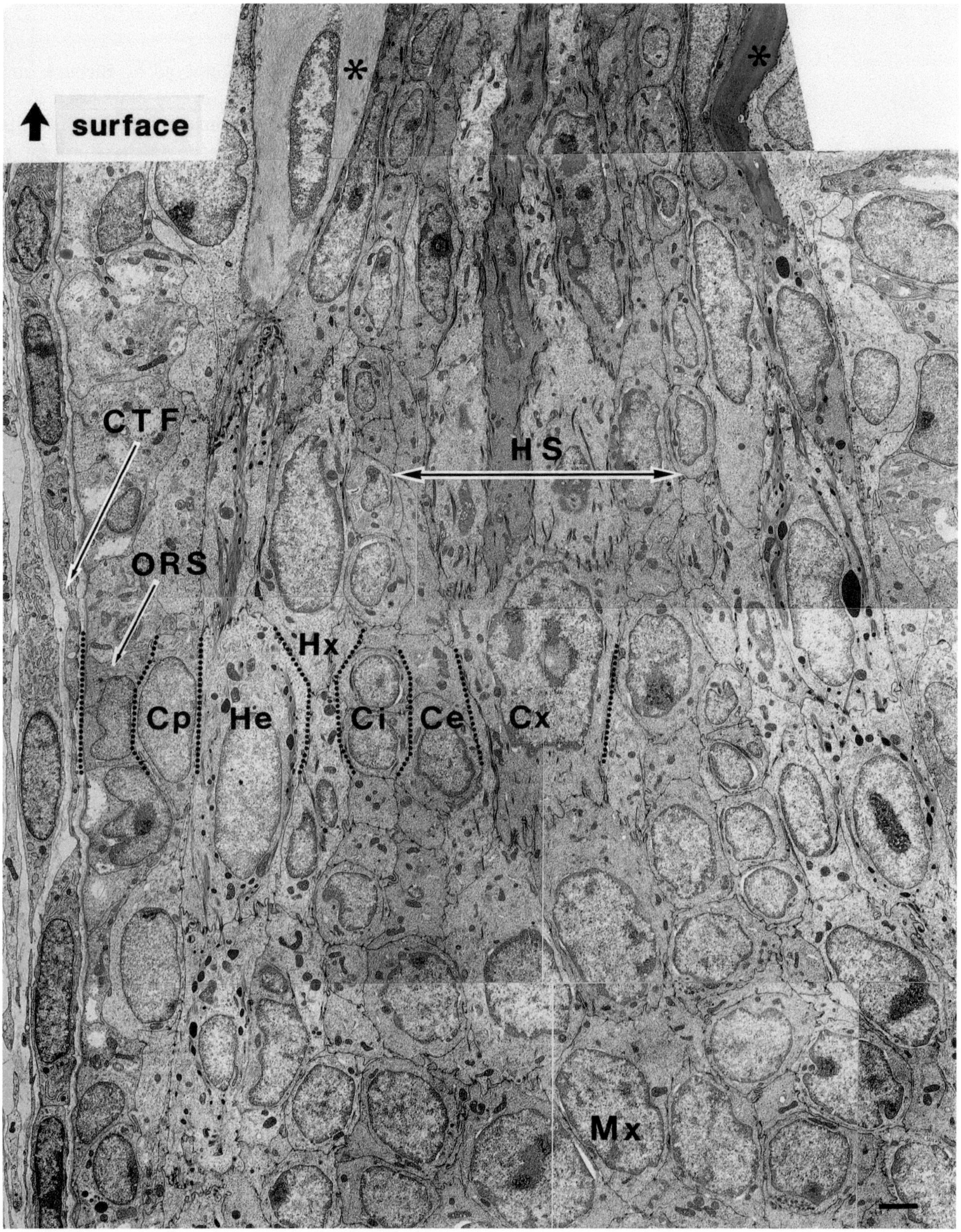

Fig. 1.5.

Electron Microscopy

Electron microscopy of the hair follicle was pioneered by Birbeck and Mercer in the middle of the last century [31–33]. Several years later, Roth and Helwig reported the ultrastructure of the mouse hair [49, 50]. While their work remains valuable, advances in apparatus and technology in the years since then should not be ignored. Herein I show the latest electron microscopic views of, first, an area slightly higher (that is, nearer to the epidermis) than the matrix (Fig. 1.5) and the equivalent at the level (of the height) of the dermal papilla (Fig. 1.7). Positions of the tissues will be confirmed by comparing these photographs with Fig. 1.4. To aid understanding, a schematic representation of the cellular boundary is shown in Fig. 1.6. The figures represent the first step to an understanding of the development of hair. Figure 1.5 shows that each tissue except the hair cortex consists of a cellular monolayer; this is probably the minimum constitution of mammalian hair.

As shown in Figs. 1.5 and 1.7, and in other figures in later sections, the cells of the connective tissue follicle (CTF) do not attach directly to those of the outer root sheath (ORS). A space of at least 0.2 μm width, which may be equivalent to the basement membrane of the skin, is observed between them. This space consists of an intercellular matrix such as collagens, laminins, fibronectins, and polysaccharides. At present an optimal terminology has not been provided for the basement membrane of hair follicles; one reason is that its biochemical specificity has not been fully elucidated. Hair follicle cells originating from epidermal ancestors adhere tightly to each other. Although most cells are not fully keratinized at the height (or stage) of Fig. 1.5, only the uppermost cells of Henle's layer (designated with asterisks) are filled with fibers, representing the earliest and most rapid keratinization of Henle's layer. The rapidity of this keratinization can be appreciated by reference to the speed of hair growth and the length of the cell, as an example, human scalp hair extends approximately 10 mm/month or 14 μm/h, meaning that the critical step of keratinization takes place within an hour, corresponding to the length of one Henle's layer cell. In other words, the critical step of the keratinization takes place during the time necessary for one-cell-length movement of the Henle's layer cells. Tonofilaments (keratin-based fibers) are also observed in other cells during the process of differentiation, such as those of Huxley's layer, the cuticle of the inner root sheath, and the hair cortex.

The short arrows in Fig. 1.6 designate the special crossroad intersections of cell surfaces located at the boundary between the Ci and Ce. It results in the integral number (practically, one or two) of Ci cells facing

Fig. 1.6. Cell boundaries in Fig. 1.5 are extracted. The boundaries are between *Ce* (hair cuticle, colored *gray*) and *Ci* (cuticle of inner root intersections), that is, approximately one or two Ci cells correspond to one Ce cell. They seem to prepare for the future special conjunctions between them, which are detailed in Chapter 4. The *patterned area* represents Cx (hair cortex)

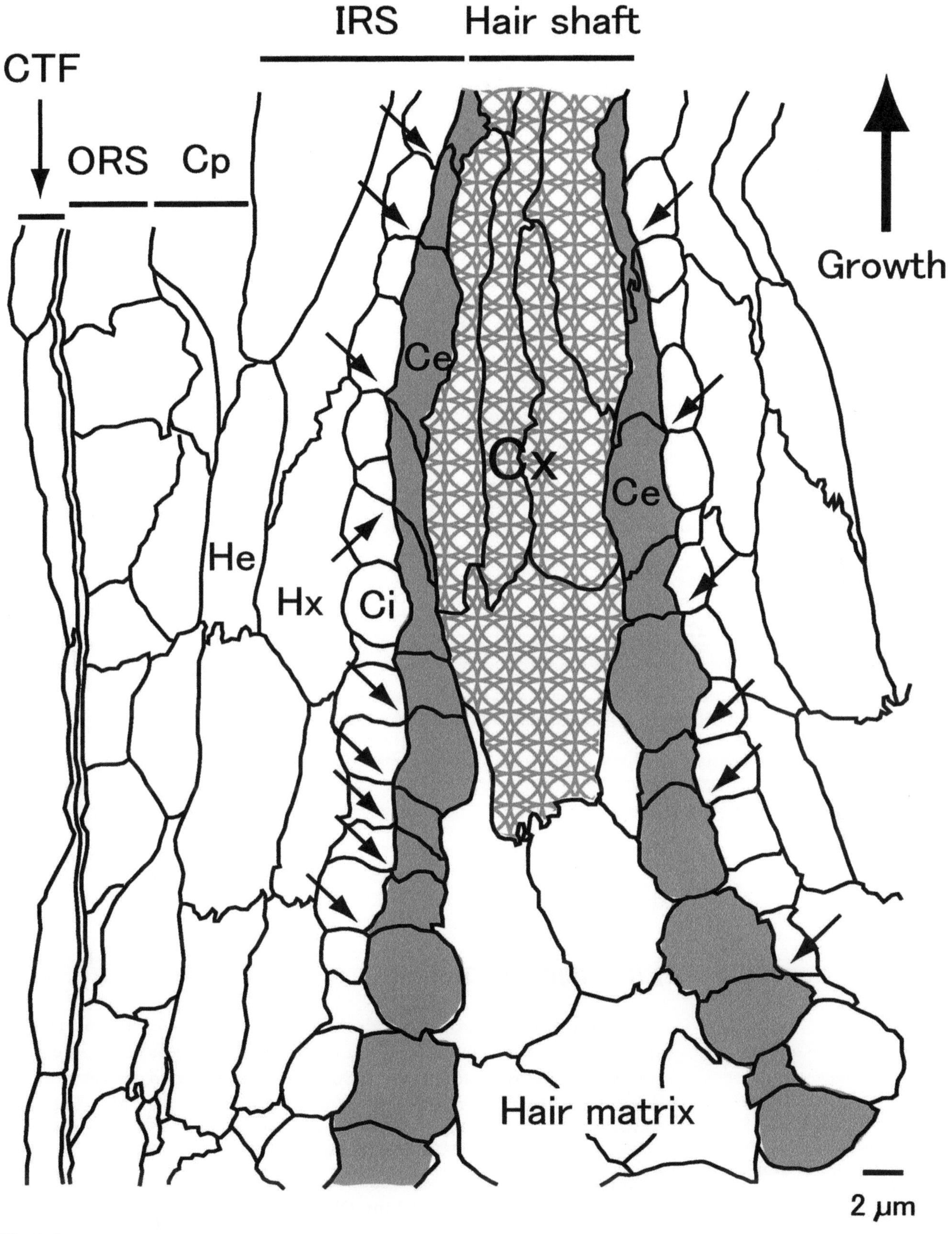
IRS
Hair shaft
CTF
ORS
Cp
Growth
Ce
Cx
Ce
He
Hx
Ci
Hair matrix
2 μm

Fig. 1.6.

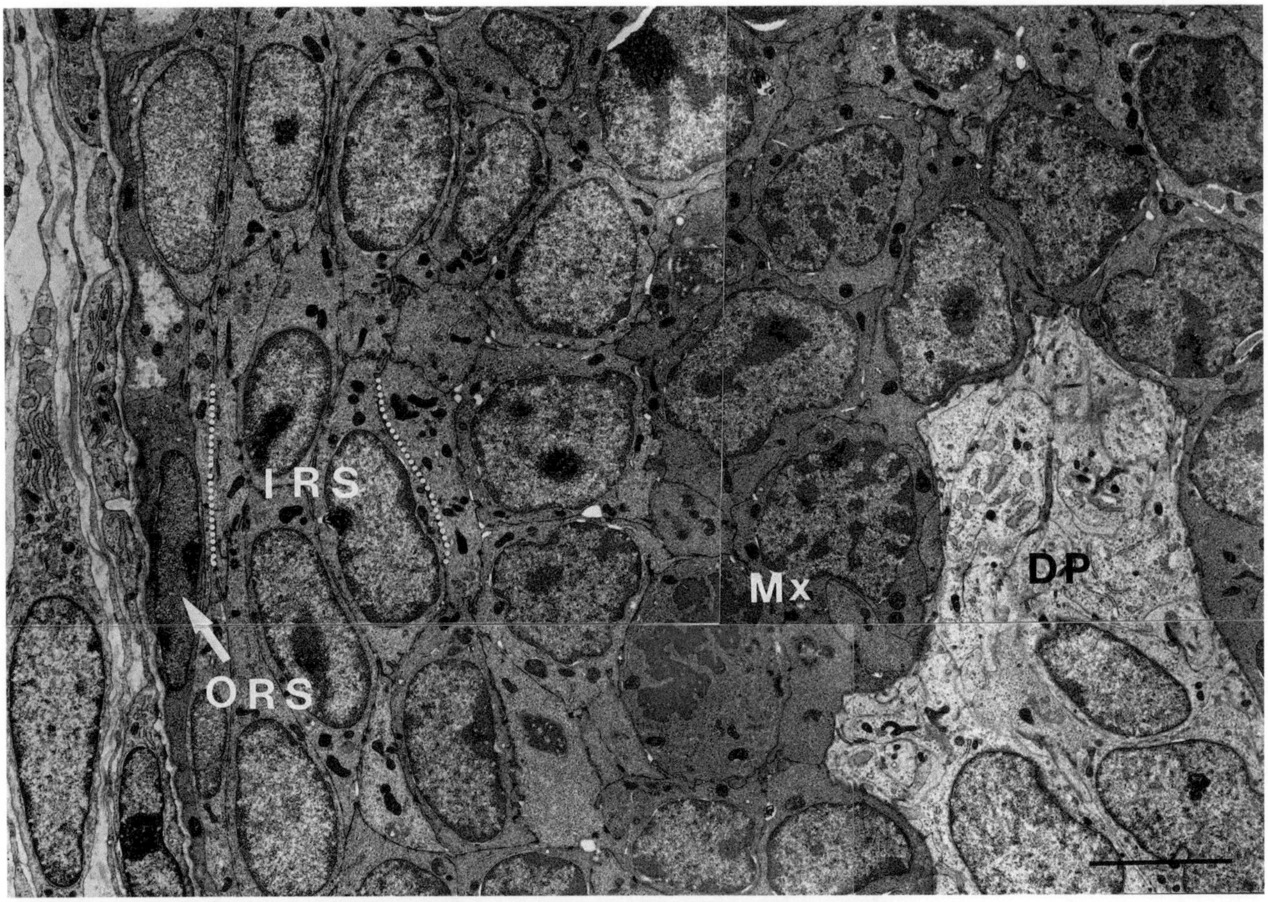

Fig. 1.7. The upper part of the dermal papilla (*DP*) and surrounding hair follicle cells. This photograph continues from the lower part of Fig. 1.5. *Dotted lines* depict the boundaries between tissues. The inner root sheath (IRS) is presumptive at this height, because its subtissues do not start to exhibit the signs of the distinctive patterns of keratinization. *ORS*, inner root sheath; *Mx*, matrix. *Bar* 5 μm

one Ce cell. Any other cell-to-cell relationships are not regulated by such a law, at least in the hair follicle. The significance of these intersections is discussed in Chapter 4.

Life and Death

At the level of the top of the dermal papilla (Fig. 1.7), the outer root sheath (ORS), the companion layer (Cp), and the three layers of the inner root sheath (IRS) are identifiable. Some undifferentiated cells around the dermal papilla (DP) appear to be undergoing cell division (Fig. 1.7). These cells are members of the hair matrix (Mx). The second layer (counted from the DP side) appears to be most active in proliferation, although statistical study has not been done. The presence of pluripotent stem cells in the area around the DP is indisputable, but the mechanism by which the various cell lineages are produced has not been elucidated. In theory both combinations, A + B and A + C, are able to form lineages (Fig. 1.8). If A + C is the operating system, every cell pool (such as a, b, and c in Fig. 1.8) must be filled rapidly by the division of pluripotent stem cells. However, it is generally believed that pluripotent stem cells do not undergo rapid cell division, and further, the root of Henle's layer, for example, may be distant from the pluripotent stem cells (located near the DP). Progenitor cells must thrust themselves successively into the cell mass to reach their assigned positions. It is therefore plausible that the system consists of processes such as the combination of A and B. In this system, a progenitor cell can perform unequal onsite cell division at the root of the specific cell lineage to produce a replica and a committed cell (horizontally striped in Fig. 1.8B). Of course, it may be possible that cell "a" divides into "a" + "a" onsite; however, the division of "a" into "a" + "another" is an essential process if a differentiated line of cells is to be

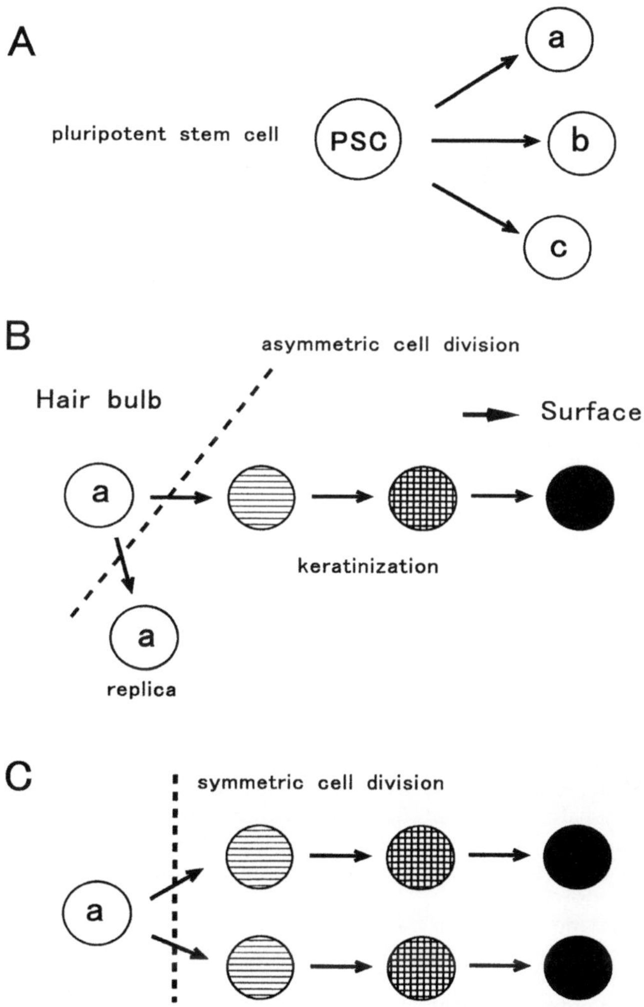

Fig. 1.8. Strategies to produce hair follicle cells. Pluripotent stem cells (*PSC*) near the dermal papilla divide and produce precursor cells of a variety of lineages (**A**). It has not been determined definitely whether those precursor cells have a potential to reproduce themselves. If they have the potential, system **B** is possible; if not, system **C** is the residual choice. The letters *a*, *b*, and *c* represent the precursor cells; *striped, meshed*, and *black circles* represent the cells undergoing terminal differentiation

produced. A comprehensive and final elucidation of the mechanism of hair growth will require the study of unequal (or asymmetrical) cell division of the hair bulb matrix cells.

The pluripotent stem cells are suggested to migrate to an upper position of the hair follicle (bulge region) before the first catagen. It is likely that they escape the degeneration of the lower follicle in the catagen. They

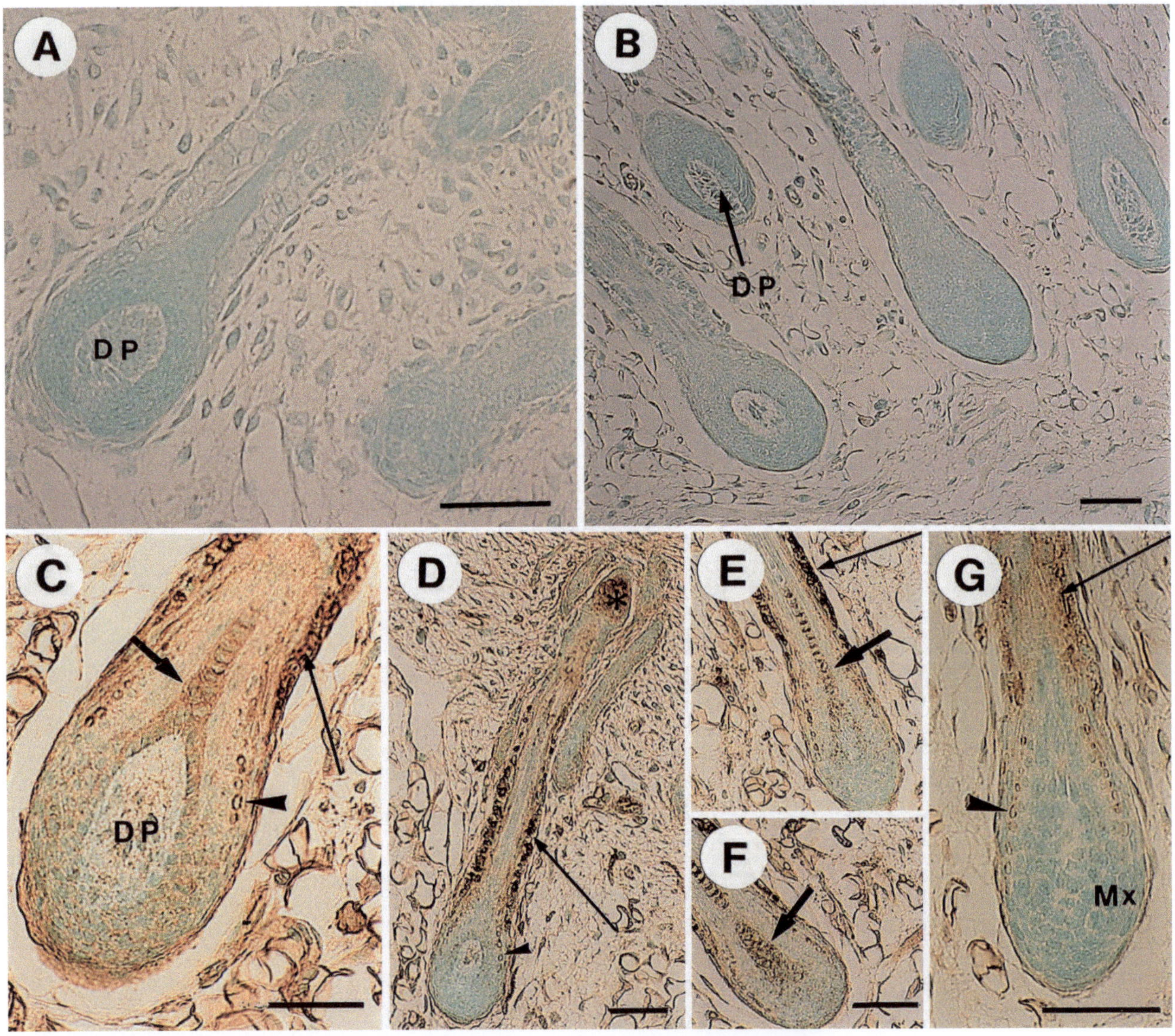

Fig. 1.9. Microscopic views of immunohistochemically stained hair follicles. **A** Follicles of neonatal rat at day 0 (day of birth). **B** Same as **A** except the sample was obtained from day 2 animals. **A** and **B** are the control experiments without treatment with primary antibodies. Samples for **C–G** were obtained from the rats at day 2 after birth. They were treated with following antibodies: **C**, anti-μ-calpain; **D–F**, anticathepsin; **B**, **G**, anticathepsin D, respectively; and were then treated with an ABC reagent kit followed by visualization with peroxidase-DAB staining system (42). Poststaining was carried out by use of methyl green. *Arrowheads*, *thick arrows*, and *thin arrows* are immunostained areas and are expected to be keratinizing Henle's layer, medulla with precursor matrix cells, and outer root sheath, respectively. The *asterisk* in **D** depicts a sebaceous gland, which is shown more fully in Fig. 1.10. *Bars* 40 μm

then return to the hair matrix to regenerate the follicle again in the next anagen [51, 52]. These pluripotent stem cells in the hair follicle have the potential to form not only most parts of the hair but also sebaceous gland and epidermis [52, 53]. I do not discuss in detail the hair cycle of adult animals or humans in this book; the subject is covered well elsewhere [54–56].

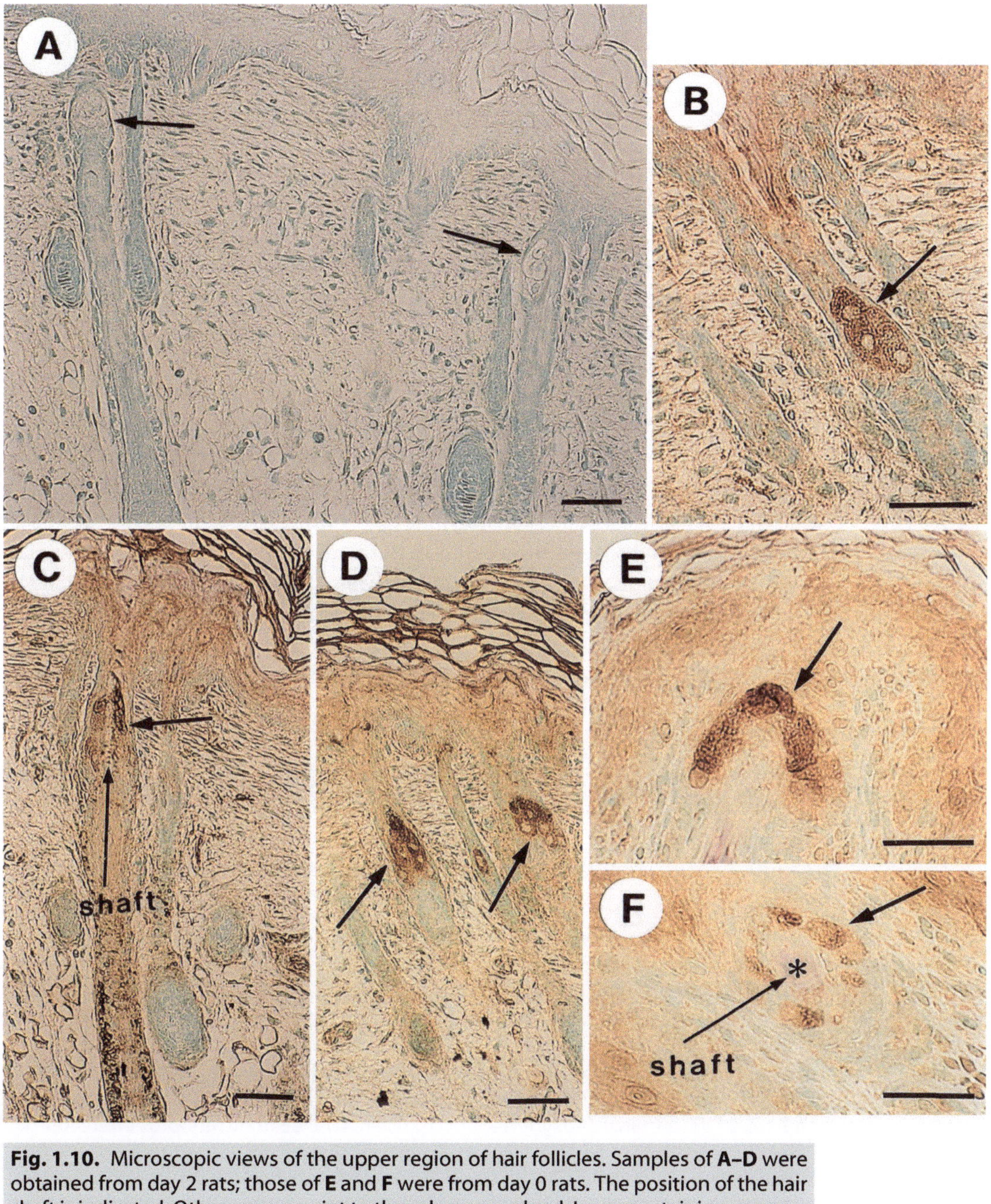

Fig. 1.10. Microscopic views of the upper region of hair follicles. Samples of **A–D** were obtained from day 2 rats; those of **E** and **F** were from day 0 rats. The position of the hair shaft is indicated. Other *arrows* point to the sebaceous gland. Immunostaining was performed by use of anticathepsin B antibody (**B**, **C**), anticathepsin D antibody (**D**), and anticathepsin L antibody (**E**, **F**), respectively. **A** is a control experiment without treatment with primary antibodies. An asterisk in F denotes a cross section of hair shaft. *Bars* 40 μm

Before going into the details of electron microscopy studies, the major subject of this book, I would like to provide some full-length microscopic views of the hair follicle and to describe some of our experimental results concerning the localization of some proteases that may contribute to the terminal differentiation of the hair follicle cells by showing the figures in our previous paper as well as unpublished results. As shown in Figs. 1.9 and 1.10, cathepsins and μ-calpain are localized in the outer root sheath of the middle and lower parts of the follicle, premature areas of Henle's layer, the medulla, and the sebaceous gland. The sebaceous gland secretes sebum to waterproof the hair and local skin. The sebaceous gland surrounds the hair shaft at an early stage (Fig. 1.10C,E,F), later localizing to one side (Fig. 1.10A,B). Except at the outer root sheath, proteases are located in those areas undergoing programmed cell death, suggesting a role for them in replacing components of the cytoplasm of these cells. The outer root sheath cells may undergo a sporadic but specific cornification in the wide area above the level of the hair bulb. The bulge region is formed at a position slightly lower than the sebaceous gland. Details of the sebaceous gland are provided in reference 58.

Morphological degeneration of the nucleus and chromatin is expected to involve the fragmentation of DNA. In particular, double-strand breaks of DNA have been used as markers of apoptosis [59–61]. An increase in the number of single-strand breaks also occurs in nonapoptotic types of cell death. Because most cells in the hair follicle undergo terminal differentiation, we have been interested in determining where and when during hair formation the fragmentation occurs. In conventional in situ DNA end-labeling assay to detect the double-strand breaks, the dermal papilla, hair matrix, and outer root sheath (or its precursor cells) stain negatively whereas the internal root sheath and hair shaft stain positively (Fig. 1.11A–F). Those positively stained areas are superimposable on the ongoing keratinizing areas. Where keratinization is complete, staining is not observed (Fig. 1.11F). I do not think that this is caused by the complete degradation of DNA, but rather by an extreme hardening of the tissue that prevents the penetration of the enzyme for the assay. Areas rich in single-strand breaks show a similar pattern of staining, suggesting that the start of keratinization involves both single- and double-strand DNA breaks (Fig. 1.11G–L). Single-strand breaks in Fig. 1.11 are not thought to be the result of the formation of Okazaki fragments because most actively dividing hair matrix cells are not stained.

Fig. 1.11. Strand breaks of DNA during the development of hair follicles. Dorsal skin of a day 0 rat was used as the material. **A–F** Double-strand breaks were detected by use of a TACS 2 TdT (terminal deoxynucleotidyl transferase)-DAB in situ apoptosis detection kit (Trevigen, Gaithersburg, MD, USA). **A**, **C**, and **E** are the controls without the enzyme, neighboring sections **B**, **D**, and **F**, respectively. **G**, **H** Single-strand breaks were detected by use of a TACS 2 Klenow (Klenow DNA polymerase)-DAB kit (Trevigen). **G**, **I**, and **K** are the controls without the enzyme, neighboring sections **H**, **J**, and **L**, respectively. Methyl green was used for the counterstaining. **A**, **B**, and **C–F** are the same magnification, respectively. *Bars* 50 μm

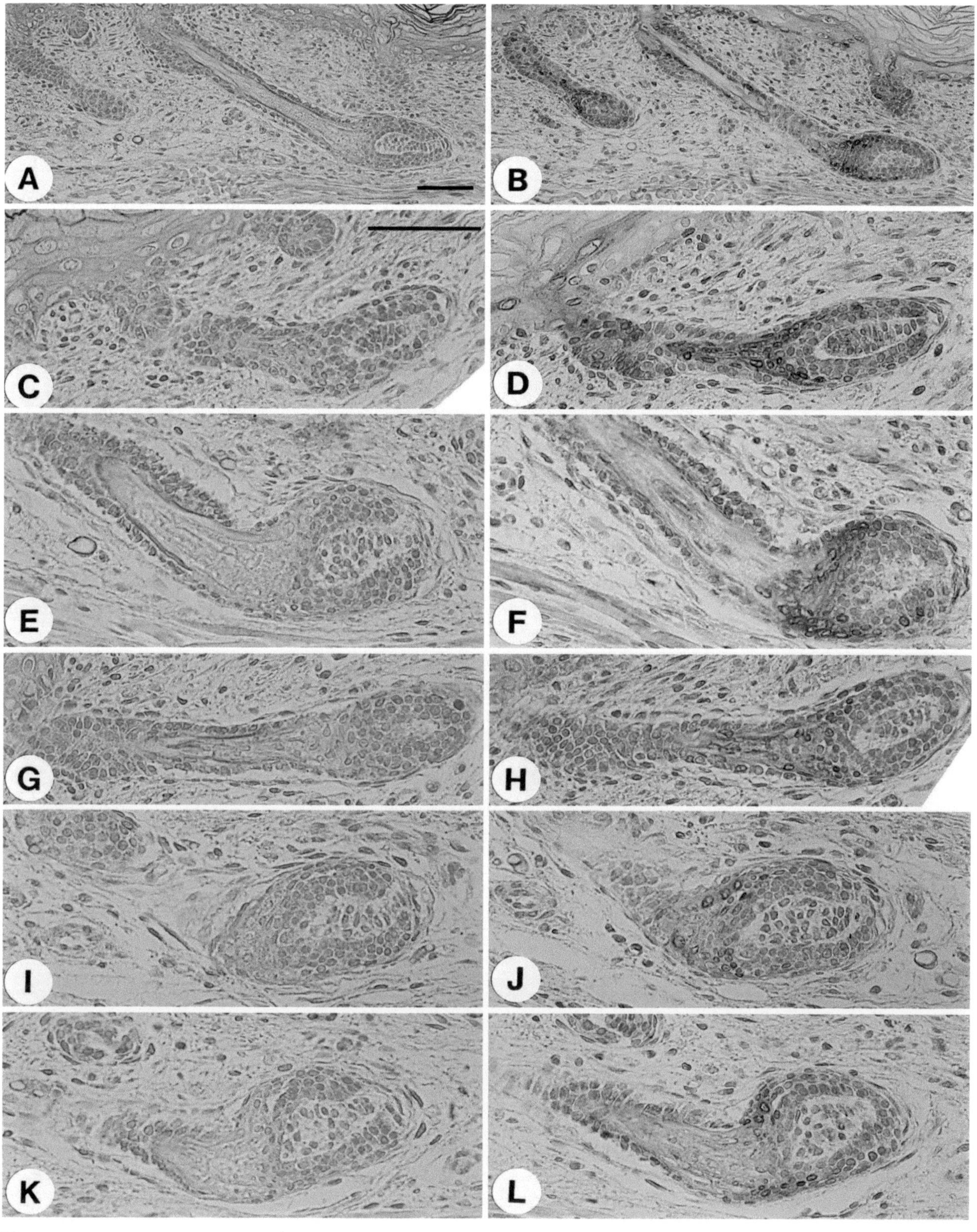

Fig. 1.11.

Materials and Methods

Here, I would like to quote the words of Swift [62]: "The cutting of these with the aid of an ultramicrotome is not an easy task (dare we say it's an art!) on account of the hair's inherent toughness and because embedding resins barely penetrate the fiber surface and do not readily key to it." I fully concur. For this reason, and because my interest is in developmental aspects of the hair, I have decided to limit myself in this book to the skin of neonates, the treatment of which is so much easier. Except for the structure of the medulla and associated organs and the location of stem cells, neonate and adult hair are essentially the same.

I have generally used conventional methods for electron microscopy. In brief, skin tissues obtained from the dorsum or face of neonatal Sprague–Dawley rats were fixed overnight in a solution of 3.7% paraformaldehyde and 1% glutaraldehyde in 50 mM sodium phosphate buffer (pH 7.2) at 4°C. They were then washed thoroughly in phosphate-buffered saline (PBS) at pH 7.2 and treated with 2% osmium tetroxide in PBS for 1 h at 4°C. Poly/Bed 812 (Epon) was used for embedding. Ultrathin sections (70–80 nm in thickness) were stained with uranyl acetate and lead citrate. The observations were carried out using a JEM 1010 electron microscope (JEOL, Tokyo, Japan). All operations including development, printing, enlargement, cutting, and patching of photographs were carried out by myself in the traditional ways without use of the computer-dependent techniques.

For immunohistochemical study, dorsal skin tissues were fixed with 10% formalin neutral buffer and stored as paraffin wax-embedded blocks. They were cut into sections as need arose. Immunoreactions were performed using a Vectastain ABC kit (PK-6200; Vector Laboratories, Burlingame, CA, USA). The reaction was visualized with a peroxidase substrate kit (SK-4100; Vector Laboratories). The cells were then post-stained with methyl green. Further experimental details are provided in references 57, 63, and 64. Sections for photomicroscopy were prepared as described by Ihara and Motobayashi [65].

References

1. Prum RO, Brush AH (2003) Hair, scales, fur, feathers. Of all body covering nature. Sci Am 288:62–69
2. Meng J, Wyss AR (1997) Multituberculate and other mammalian hair recovered from Palaeocene excreta. Nature (Lond) 385:712–714
3. Dykes T. Mesozoic Eucynodonts: genus index. http://home.arcor.d/ktdykes/meseucaz.htm
4. Savage RJG, Long MR (1986) Mammal evolution: an illustrated guide. British Museum, London
5. Knafelc KA (2000) From reptiles to mammals: evolution of early mammals. The Compass Rose 1:1–5. http://www.dragonlordsnet.com/erfrm.htm
6. Kaneko R (1998) Mammalian reptiles (in Japanese). Asahi Press, Tokyo
7. Linder G, Botchkarev VA, Botchkareva NV, Ling G, van der Veen C, Paus R (1997) Analysis of apoptosis during hair follicle regression (catagen). Am J Pathol 151:1601–1617
8. Paus R, Cotsarelis G (1999) The biology of hair follicles. N Engl J Med 341:491–497
9. Oshima H, Rochat A, Kedzia C, Kobayashi K, Barrandon Y (2001) Morphogenesis and renewal of hair follicles from adult multipotent stem cells. Cell 104:233–245
10. Paus R, Muller-Rover S, Botchkarev VA (1999) Chronobiology of the hair follicle: hunting the "hair cycle clock." JID Symp Proc 4:338–345
11. Hardy MH (1992) The secret life of the hair follicle. Trends Genet 2:55–61
12. Chuong CM, Chodankar R, Widelitz RVB, Jiang TX (2000) Evo-devo of feathers and scales: building complex epithelial appendages. Curr Opin Dev Genet 10:449–456
13. Dhouailly D, Prin F, Kanzler B, Viallet JP (1998) Variations of cutaneous appendages: regional specification and cross-species signals. In: Chuong CM (ed) Molecular basis of epithelial appendage morphogenesis. Landes Bioscience, Austin, TX, pp 45–56
14. Cadigan KM, Nusse R (1997) Wnt signaling: a common theme in animal development. Genes Dev 11:3286–3305
15. Gat U, DasGupta R, Degenstein L, Fuchs E (1998) De novo hair follicle morphogenesis and hair tumors in mice expressing a truncated β-catenin in skin. Cell 95:605–614
16. Millar SE, Willert K, Salinas PC, Roelink H, Nusse R, Sussman DJ, Barsh GS (1999) Wnt signaling in the control of hair growth and structure. Dev Biol 207:133–149
17. Hung BS, Wang X-Q, Cam GR, Rothnagel JA (2001) Characterization of mouse Frizzled-3 expression in hair follicle development and identification of the human homologue in keratinocytes. J Invest Dermatol 116:940–946
18. Huelsken J, Vogel R, Erdmann B, Costarelis G, Birchmeier W (2001) β-Catenin controls hair follicle morphogenesis and stem cell differentiation in the skin. Cell 105:533–545
19. van Genderen C, Okamura RM, Farinas I, Quo R-G, Parslow TG, Bruhn L, Grosschedl R (1994) Development of several organs that require inductive epithelial-mesenchymal interactions is impaired in LEF-1-deficient mice. Genes Dev 8:2691–2703
20. Jamora C, Das Gupta R, Kocieniewski P, Fuchs E (2003) Links between signal transduction, transcription and adhesion in epithelial bud development. Nature (Lond) 422:317–322
21. Van Mater D, Kollings FT, Dlugosz AA, Fearon ER (2003) Transient activation of beta-catenin signaling in cutaneous keratinocytes is sufficient to trigger the active growth phase of the hair cycle in mice. Genes Dev 17:1219–1224
22. Botchkarev VA, Botchkareva NV, Nakamura M, Huber O, Funa K, Lauster R, Paus R, Gilchrest BA (2001) Noggin is required for induction of the hair follicle growth phase in postnatal skin. FASEB J 15:2205–2214
23. Botchkarev VA (2003) Bone morphogenetic proteins and their antagonists in skin and hair follicle biology. J Invest Dermatol 120:36–47
24. Powell BC, Passmore EA, Nesci A, Dunn SM (1998) The Notch signaling pathway in hair growth. Mech Dev 78:189–192
25. Favier B, Fliniaux I, Thelu J, Viallet JP, Demarchez M, Jahoda CAB, Dhouailly D (2000) Localisation of members of the Notch system and the differentiation of vibrissa hair follicles: receptors, ligands, and fringe modulators. Dev Dyn 218:426–437
26. Lin M-H, Leimeister C, Gessler M, Kopan R (2000) Activation of the Notch pathway in the hair cortex leads to aberrant differentiation of the adjacent hair-shaft layers. Development (Camb) 127:2421–2432
27. Foitzik K, Paus R, Doetschman T, Dotto GP (1999) The TGF-β2 isoform is both a required and sufficient inducer of murine hair follicle morphogenesis. Dev Biol 212:278–289
28. Headon DJ, Overbeek PA (1999) Involvement of a novel Tnf receptor homologue in hair follicle induction. Nat Genet 22:370–374
29. Alonso L, Fuchs E (2003) Stem cells in the skin: waste not, Wnt not. Genes Dev 17:1189–1200
30. Fuchs E, Raghavan S (2002) Getting under the skin of epidermal morphogenesis. Nat Rev Genet 3: 199–209
31. Birbeck MSC, Mercer EH (1957) The electron microscopy of the human hair follicle. I:

Introduction and the hair cortex. J Biophys Biochem Cytol 3:203–214
32. Birbeck MSC, Mercer EH (1957) The electron microscopy of the human hair follicle. II: The hair cuticle. J Biophys Biochem Cytol 3:215–221
33. Birbeck MSC, Mercer EH (1957) The electron microscopy of the human hair follicle. III: The inner root sheath and trichohyaline. J Biophys Biochem Cytol 3:223–229
34. McGowan KM, Coulombe PA (2000) Keratin 17 expression in the hard epithelial context of the hair and nail, and its relevance for the pachyonychia congenital phenotype. J Invest Dermatol 114:1101–1107
35. Heid HW, Moll I, Franke WW (1988) Patterns of expression of trichocytic and epithelial cytokeratins in mammalian tissues. I. Human and bovine hair follicles. Differentiation 37:137–157
36. Langbein L, Rogers MA, Winter H, Praetzel S, Beckhaus U, Rackwitz H-R, Schweizer J (1999) The catalog of human hair keratins. I. Expression of the nine type I members in the hair follicle. J Biol Chem 274:19874–19884
37. Langbein L, Rogers MA, Winter H, Praetzel S, Schweizer J (2001) The catalog of human hair keratins. II. Expression of the six type II members in the hair follicle and the combined catalog of human type I and type II keratins. J Biol Chem 276:35123–35132
38. Heid HW, Moll I, Franke WW (1988) Patterns of expression of trichocytic and epithelial cytokeratins in mammalian tissues. II. Concomitant and mutually exclusive synthesis of trichocytic and epithelial cytokeratins in diverse human and bovine tissues (hair follicle, nail bed and matrix, lingual papilla, thymic reticulum). Differentiation 37:215–230
39. Stark HJ, Breitkreutz D, Limat A, Ryle CM, Roop D, Leigh I, Fusenig N (1990) Keratins 1 and 10 or homologues as regular constituents of inner root sheath and cuticle cells in the human hair follicle. Eur J Cell Biol 52:359–372
40. Fuchs E (1996) The cytoskeleton and disease: genetic disorders of intermediate filaments. Annu Rev Genet 30:197–231
41. Schirren CG, Burgdorf WH, Sander CA, Plewig G (1997) Fetal and adult hair follicle. An immunohistochemical study of anticytokeratin antibodies in formalin-fixed and paraffin-embedded tissue. Am J Dermatopathol 19:335–340
42. Bawden CS, McLaughlan C, Nesci A, Rogers G (2001) A unique type I keratin intermediate filament gene family is abundantly expressed in the inner root sheaths of sheep and human hair follicles. J Invest Dermatol 116:157–166
43. Langbein L, Rogers MA, Praetzel S, Winter H, Schweizer J (2003) K6irs1, K6irs2, K6irs3, and K6irs4 represent the inner-root-sheath-specific type II epithelial keratins of the human hair follicle. J Invest Dermatol 120:512–522
44. Bernot KM, Coulombe PA, McGowan KM (2002) Keratin 16 expression defines a subset of epithelial cells during skin morphogenesis and the hair cycle. J Invest Dermatol 119:1137–1149
45. Winter H, Langbein L, Praetzel S, Jacobs M, Rogers MA, Leigh IM, Tidman N, Schweizer J (1998) A novel human type II cytokeratin, K6hf, specifically expressed in the companion layer of the hair follicle. J Invest Dermatol 111:955–962
46. Whitbread LA, Powell BC (1998) Expression of the intermediate filament keratin gene, K15, in the basal cell layers of epithelia and the hair follicle. Exp Cell Res 244:448–459
47. Wang Z, Wong P, Langbein L, Schweizer J, Coulombe PA (2003) Type II epithelial keratin 6hf (K6hf) is expressed in the companion layer, matrix, and medulla in anagen-stage hair follicles. J Invest Dermatol 121:1276–1282
48. Orwin DFG (1971) Cell differentiation in the lower outer sheath of the Romney wool follicle. A companion cell layer. Aust J Biol Sci 24:989–999
49. Roth SI, Helwig EB (1964) The cytology of the dermal papilla, the bulb, and the root sheaths of the mouse hair. J Ultrastruct Res 11:33–51
50. Roth SI, Helwig EB (1964) The cytology of the cuticle of the cortex, the cortex, and the medulla of the mouse hair. J Ultrastruct Res 11:52–67
51. Rochat A, Kobayashi K, Barrandon Y (1994) Location of stem cells of human hair follicles by clonal analysis. Cell 76:1063–1073
52. Oshima H, Rochat A, Kedzia C, Kobayashi K, Barrandon Y (2001) Morphogenesis and renewal of hair follicles from adult multipotent stem cells. Cell 104:233–245
53. Taylor G, Lehrer MS, Jensen PJ, Sun T-T, Lavker RM (2000) Involvement of follicular stem cells in forming not only the follicle but also the epidermis. Cell 102:451–461
54. Young RD, Oliver RF (1976) Morphological changes associated with the growth cycle of vibrissal follicles in the rat. J Embryol Exp Morphol 36:597–607
55. Paus R, Cotsarelis G (1999) The biology of hair follicles. N Engl J Med 341:491–497
56. Muller-Rover S, Handjiski B, van der Veen C, Eichmuller S, Foitzik K, McKay IA, Stenn KS, Paus R (2001) A comprehensive guide for the accurate classification of murine hair follicles in distinct hair cycle stages. J Invest Dermatol 117:3–15
57. Morioka K, Sato-Kusubata K, Kawashima S, Ueno T, Kominami E, Sakuraba H, Ihara S (2001) Localization of cathepsins B, D, L, LAMP-1 and μ-calpain in developing hair follicles. Acta Histochem Cytochem 34:337–347

58. Thody AJ, Shuster S (1989) Control and function of sebaceous glands. Physiol Rev 69:383–416
59. Gavrieli Y, Sherman Y, Ben-Sasson SA (1992) Identification of programmed cell death in skin via specific labeling of nuclear DNA fragmen-tation. J Cell Biol 119:493–501
60. Wyllie AH (1980) Glucocorticoid-induced thymocyte apoptosis is associated with endogenous endonuclease activation. Nature (Lond) 284:555–556
61. Morioka K, Tone S, Mukaida M, Takano-Ohmuro H (1998) The apoptotic and nonapoptotic nature of the terminal differentiation of erythroid cells. Exp Cell Res 240:206–217
62. Swift JA (1997) Morphology and histochemistry of human hair. In: Jolles P, Zahn H, Hocker H (eds) EXS: Formation and structure of human hair. Birkhauser, Basel pp 149–175
63. Morioka K, Takano-Ohmuro H, Sameshima M, Ueno T, Kominami E, Sakuraba H, Ihara S (1999) Extinction of organelles in differentiating epidermis. Acta Histochem Cytochem 32:465–476
64. Takano-Ohmuro H, Mukaida M, Kominami E, Morioka K (2000) Autophagy in embryonic erythroid cells: its role in maturation. Eur J Cell Biol 79:759–764
65. Ihara S, Motobayashi Y (1992) Wound closure in foetal rat skin. Development 114:573–582
66. Celso CL, Prowse DM, Watt FM (2004) Transient activation of β-catenin signalling in adult mouse epidermis is sufficient to induce new hair follicles but continuous activation is required to maintain hair follicle tumors. Development 131:1787–1799
67. Yuhki M, Yamada M, Kawano M, Iwasato T, Itohara S, Yoshida H, Ogawa M, Mishina Y (2004) BMPRIA signaling is necessary for hair follicle cycling and hair shaft differentiation in mice. Development 131:1825–1833

Medulla 2

Variety

The medulla (abbreviated as M in the figures) is the most variable tissue in the hair follicle; excluding curl and color, the nature of the hair is considerably dependent on it. Although medulation occurs in most mammals [1], it is at the least absent in fine wool fibers [2] and pelage of the polar bear [3]. Hisa described the hairs of the polar bear as "end-capped straw" [3]. According to a textbook of forensic medicine [4], ordinary human and primate hair have discontinuous medullae. The medulla may not be detectable in lanugos of some mammals, including human babies. The degree of medulation varies in ethnic groups also. Head hairs in Caucasians (especially blond hair) often have no medulla [5, 6]. Limb hair is less medulated than scalp hair [7], while all male beard hair is medulated [6]. Cross sections of pig hairs obtained from both adults and embryos that appeared in a textbook of histology [8] do not include a medulla, whereas hairs of the adult rabbit, mouse, dog, and horse contain well-developed medulla [4, 9].

It is noteworthy that the growth and decay of the medulla is greatly variable from species to species (even among those closely related), from fetus to adult, from hair to hair in the same individual, and from one area to another in the same hair shaft. This variation may have physiological significance. For example, the large cavities in the hairs of the polar bear mentioned earlier may play a role in maintaining body temperature under cold conditions, sometimes lower than minus 50°C. I expect that some missing medullas were initially produced but thereafter degenerated as the hair grew.

In most cases, the medulla is formed as the center core of the hair and is hardened not by keratinization but by isopeptide bond-mediated intercalations of trichohyalin molecules [10]. Trichohyalin is one of the major protein components in both the internal root sheath and medulla. Some arginine residues of the trichohyalin molecule are enzymatically transferred into citrulline to produce the intercalations. It is possible to assume that the medulla is a vestigial organ and that some Mesozoic or Paleozoic reptiles might have had hairs that consisted of medulla alone. Assuming that they had no cuticle, this would hamper the identification of hairs in fossils. Another possibility is that the medulla appeared later as a response to the coldness of the glacial period.

Early Stages

I have studied the fate of the medulla in Sprague–Dawley rats from the day of birth (day 0) to 2 weeks after birth. At least some lanugos grow medullae. The immature medulla with surrounding follicle tissues is shown in Fig. 2.1. In this figure, the hair grows from left to right. The numbers attached to the cells represent the order of maturation. Number 1 is the youngest, with a few very small trichohyalin granules of diameters less than 0.1 μm. In contrast, cells 2 to 4 have granules of diameters greater than 0.2 μm. Trichohyalin granules are also observed in Henle's (He) and Huxley's (Hx) layers [they also appear in the cuticle of the inner root sheath (Ci) layer in the later stages of development]. These layers (He, Hx and Ci) belonging to the inner root sheath are filled with tonofilaments that consist of keratins and associated proteins, whereas the medulla does not produce well-developed keratin-based filaments. As shown in Fig. 2.1, vacuoles occur only in the cells of the medulla; no other layers contain vacuoles. Another point of the interest is the difference between the medullary cells and the others in their arrangement or shape: the medullary cells are piled up on the hair matrix like flat cushions (from left to right in Fig. 2.1), while all others (from the Cx layer to He layer in Fig. 2.1) are extended in the bulb-to-apex direction along the longitudinal axis.

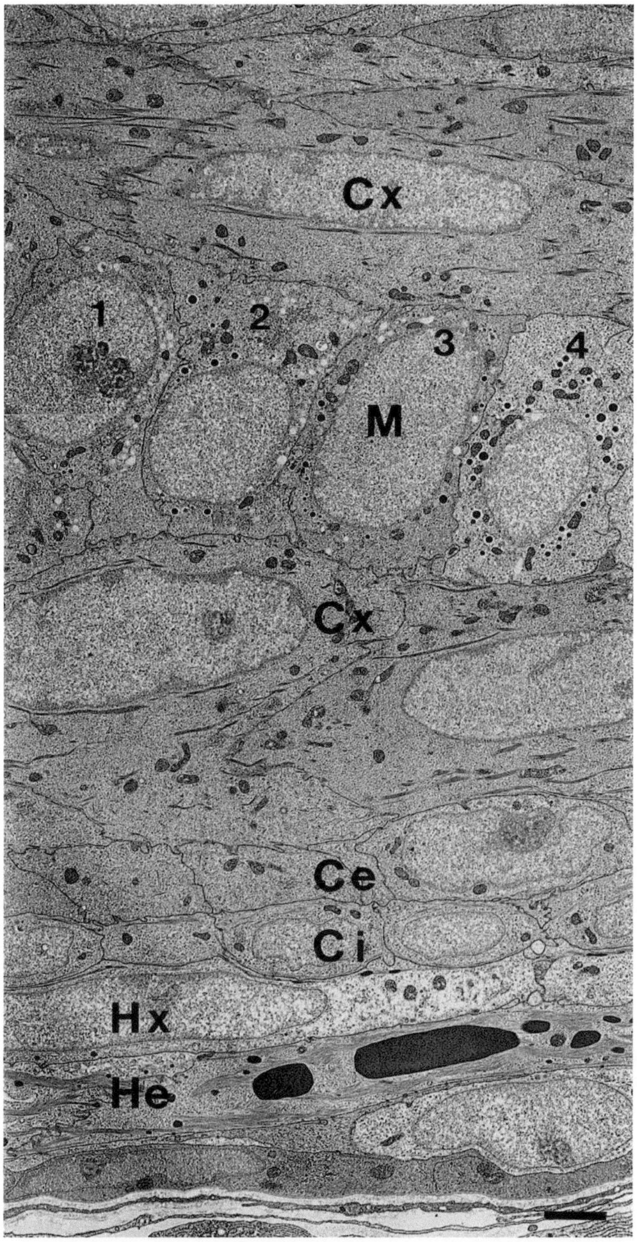

Fig. 2.1. Immature medulla and surrounding hair follicle tissues. *M*, medulla; *Cx*, hair cortex; *Ce*, hair cuticle; *Ci*, cuticle of inner root sheath; *Hx*, Huxley's layer; *He*, Henle's layer. Numbers *1–4* assigned to medulla cells suggest the order of cellular maturity. Hair grows from *left* to *right*. Dorsal skin of a day 7 rat was used as the material. *Bar* 2 μm

The dermal papilla (DP), initially spherical, becomes tubelike as the follicle matures (Fig. 2.2). Maturation of the DP is described in detail later. In matured follicles, medulla cells often start to differentiate at the side areas (not the top) of the DP, as shown in Fig. 2.2A. The small size of granules (identified by arrows in the cells numbered 1 in Fig. 2.2A and 2 in Fig. 2.2A as well as 2.2C) and that they contain no large granules suggest that they are immature and have probably just begun to differentiate toward mature medulla. Cell number 1 in Fig. 2.2C, which displays no granules, may be an early progenitor of medulla cells. Differentiation is not necessarily synchronous in cells of the same height, because neighboring cells sometimes exhibit dissimilar stages with regard to the size and frequency of trichohyalin granules in their cytoplasm (Fig. 2.2B). This finding may imply that the production and growth of the granules is rapid, occurring within several hours or so.

Trichohyalin

The mechanism of the association of trichohyalin molecules was studied by Rogers and colleagues [11]. They showed that up to 90% of the trichohyalin arginine residues are converted to citrulline residues by the action of peptidylarginine deiminase during the differentiation of the inner root sheath and medulla cells, and that a transglutaminase catalyzes the formation of γ-peptide bonds, linking glutamil and lysil residues [10, 11]. They concluded that it is undoubtedly from the γ-peptide cross-linking that the insolubility of the intracellular contents of both mature medulla and IRS is derived. They subsequently analyzed the gene and found that sheep trichohyalin has a molecular weight of 201,172 daltons, and is predicted to form an elongated alpha-helical rod structure but not to contain the sequences required for the formation of intermediate filaments [12]. The DNA sequence of the human trichohyalin gene has also been published [13], showing EF handlike calcium-binding sites* suggestive of as yet unknown functions of trichohyalin, in addition to its role as a structural protein of cornified tissues.

* EF hand is a molecular structure consists of helix-loop-helix in the protein. The loop often binds a calcium ion. Some proteins having this structure are known to be calcium signal transducers.

Unique Exocytosis

In the course of studying trichohyalin granules in the medulla, I was conscious of the fact that, in addition to this location, some are also seen in the hair cortex (Cx) (Fig. 2.3). As it is believed that the hair cortex is keratinized without the presence of keratohyalin (trichohyalin) granules [14], this puzzled me. Subsequent search of the literature revealed that Ito and Hashimoto had suggested the occasional appearance of trichohyalin

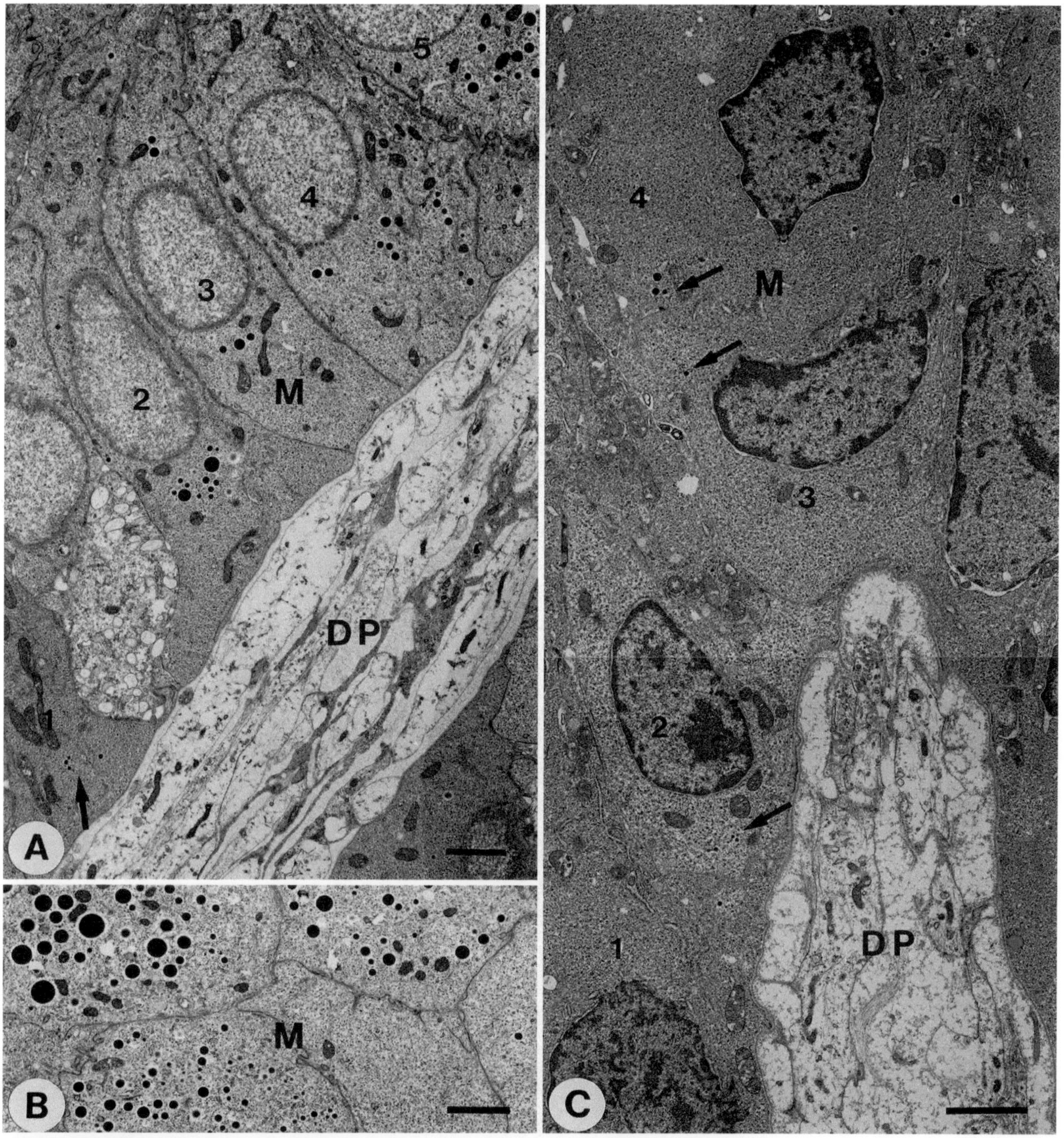

Fig. 2.2. Immature medulla cells in the vicinity of tube-like dermal papilla. *M*, medulla; *DP*, dermal papilla. **A**, **B** Dorsal skin of day 7 rats. **C** Facial skin of a day 14 rat. *Arrows* designate juvenile trichohyalin granules. *Bars* 2 μm

granules in hair cortex cells in humans and guinea pigs, but they could not find them in mice [15]. The origin and physiological meaning of these granules are not known.

The granules identified by arrows or arrowheads in Fig. 2.3A–C are apparently present in the cytoplasm of the cortex cells; they were especially frequently observed in the vicinity of the medulla. The granule denoted by an arrow in Fig. 2.3D (as well as those in other panels in Figs. 1.4, 2.4, and 2.5, to be mentioned later) suggests exocytotic degranulation from medulla to cortex. Exocytosis of granules is known to occur in mast cells stimulated by antibody and various drugs [16], but in these cases it represents a transfer from a cell to the intercellular matrix. In the present case, granules are transferred directly from one cell to another. This process involves simultaneous and coordinated exocytosis and endocytosis in adjacent cells—in other words, the cell-to-cell transfer of cytoplasm. It is not a rare event in the hair, as shown in Fig 2.3, although it seems to occur for a only limited time during the course of differentiation of the medullary cells. As far as I am aware, this type of exocytosis has not been previously reported.

Each medullary cell has two kinds of neighbor, other medulla cells and cortex cells. "Cell-to-cell transfer of the cytoplasm" seems to occur only at the fixed polarity from medulla to cortex; that is, cytoplasmic components of the medulla are transferred to the cytoplasm of hair cortex cells (Figs. 2.4–2.6). The sections of the protruded tubes of cell membrane are ob-

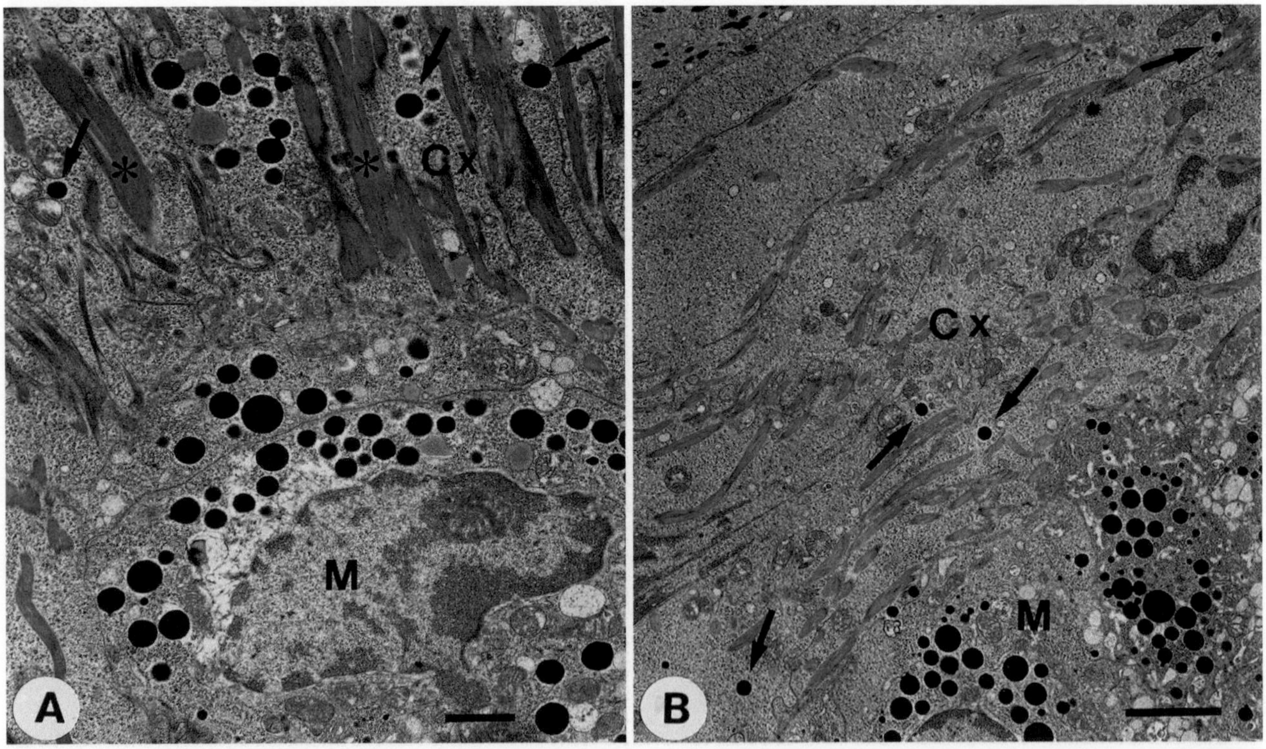

Fig. 2.3.

servable only in the cortex cells but are not found in the medulla cell (Figs. 2.4–2.6). This process is at least of nutritional benefit to the cortex, and may also involve further operations such as the transfer of regulatory factors. Although granules are extruded together with other cytoplasmic components, it is unlikely that the degranulation itself has any specific physiological role, as tonofilament formation in the hair cortex occurs independently of trichohyalin granules [14]. Because the dermal papilla of neonatal rats has not been vascularized, nutritional support by this peculiar system may be vital.

The cell-to-cell transfer occurs at the near-center position between the two arrows and never around the medulla cell-to-medulla cell junctures (shown schematically in Fig. 2.5). Active areas of cytoplasmic transfer often occur bilaterally, as designated by arrows 1 and 2 in Fig. 2.4C and shown schematically in Fig. 2.5. The site designated by arrows in Fig. 2.6B suggests that the area of exocytosis forms a ringlike structure that surrounds the medullary cell as the three-dimensional (3-D) image. The exocytosis in Fig 2.6 is active at the boundary designated by the black arrow but not at the side designated by the open arrow in the ring. This difference suggests that the active area is restricted, or is transient if mobile. It is not certain that Figs. 2.3A and 2.6A represent the top of the medulla tissue, and they may actually be oblique views; nevertheless, it remains possible that the top area is active in the transfer (see arrow in Fig. 2.6A).

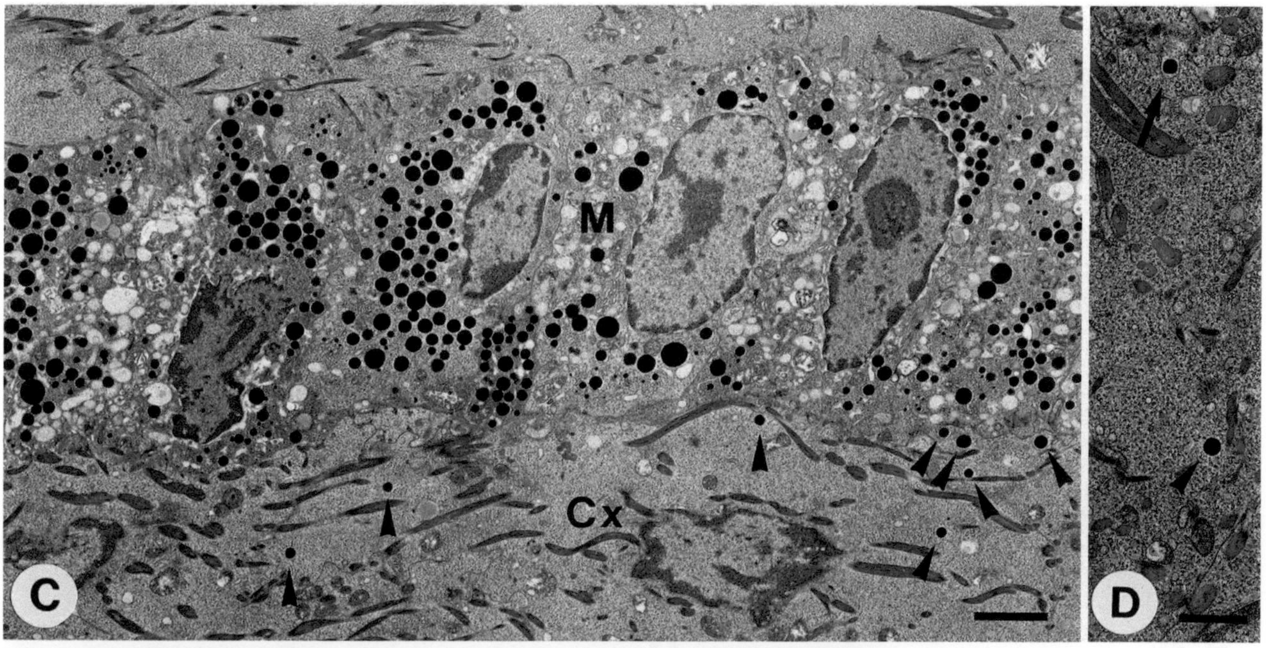

Fig. 2.3. Transfer of trichohyalin granules from medulla to hair cortex. *M*, medulla; *Cx*, hair cortex. Samples were obtained from dorsal skin of day 14 rats. *Arrows* and *arrowheads* designate the transferred granules. Most of the granules in the cortex are found in the vicinity of the boundary between the cortex and the medulla. *Asterisks* in **A** designate keratin fiber bundles in the cortex. *Bars* **A, D** 1 μm; **B, C** 2 μm

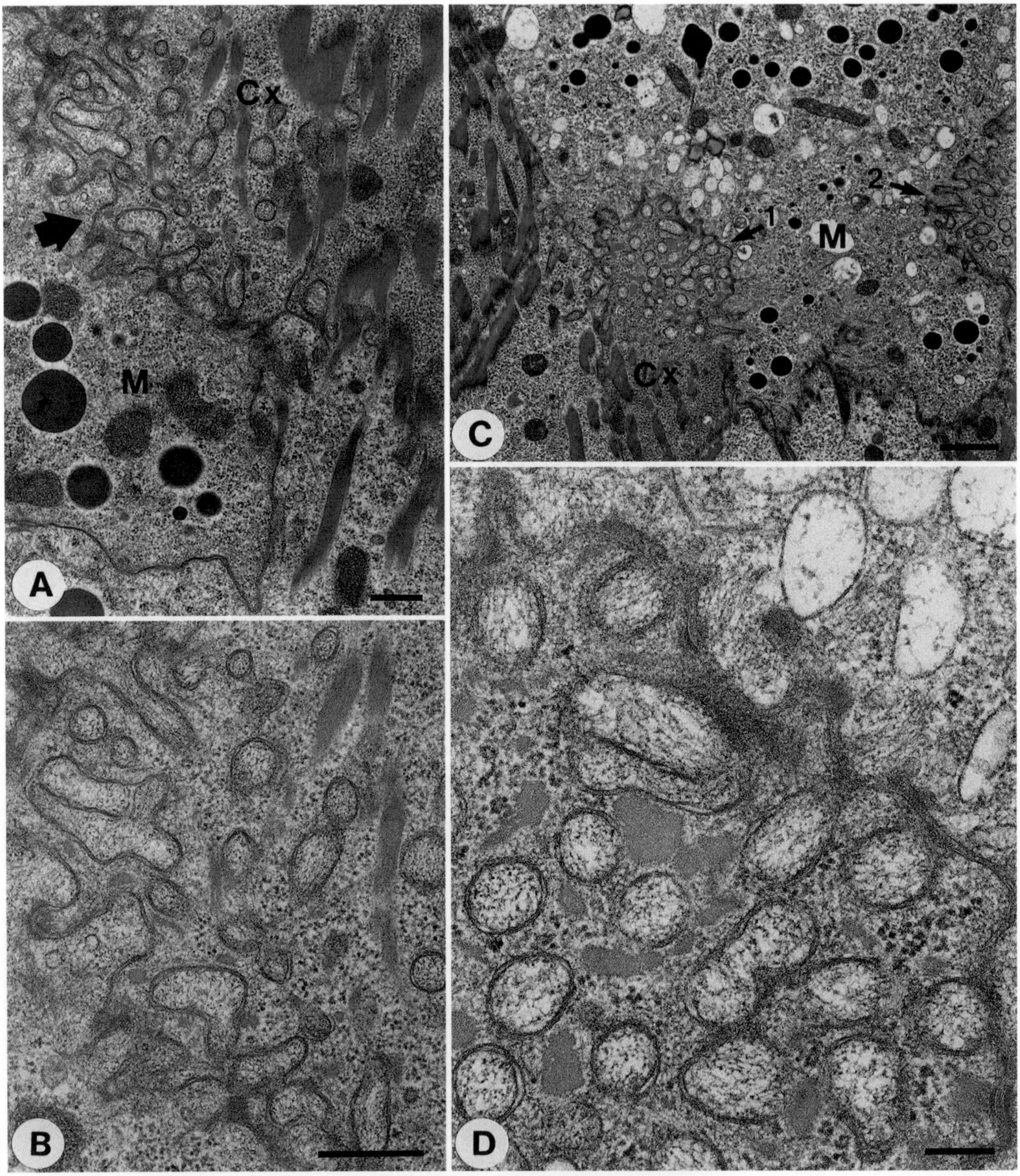

Fig. 2.4. Unique structure of the boundaries between medulla cells and cortex cells. *M*, medulla; *Cx*, hair cortex. *Arrows* represent the positions of pseudopodia formation or membrane bubbling. *Arrows 1* and *2* in **C** show a symmetrical emergence of this unique movement of the membrane. Day 7 rats were used for the experiment. *Bars* **A**, **B** 500 nm; **C** 1 μm; **D** 200 nm

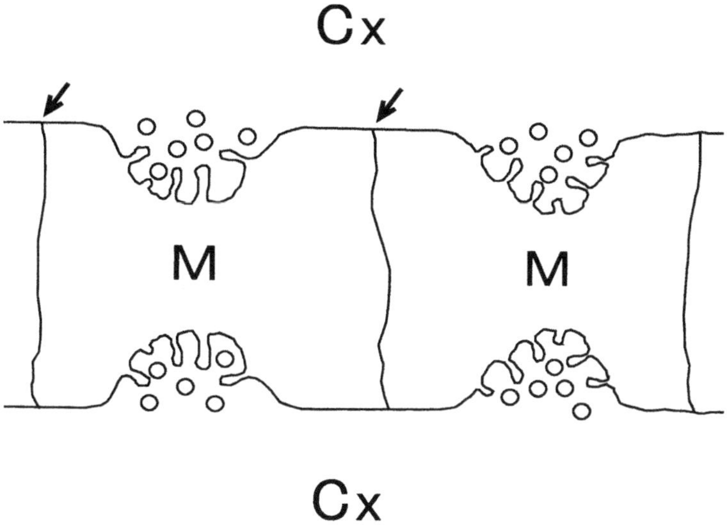

Fig. 2.5. Schematic presentation of transfer of the cytoplasm from medulla to cortex. *M*, medulla; *Cx*, hair cortex. *Arrows* indicate the junctures of cell membranes

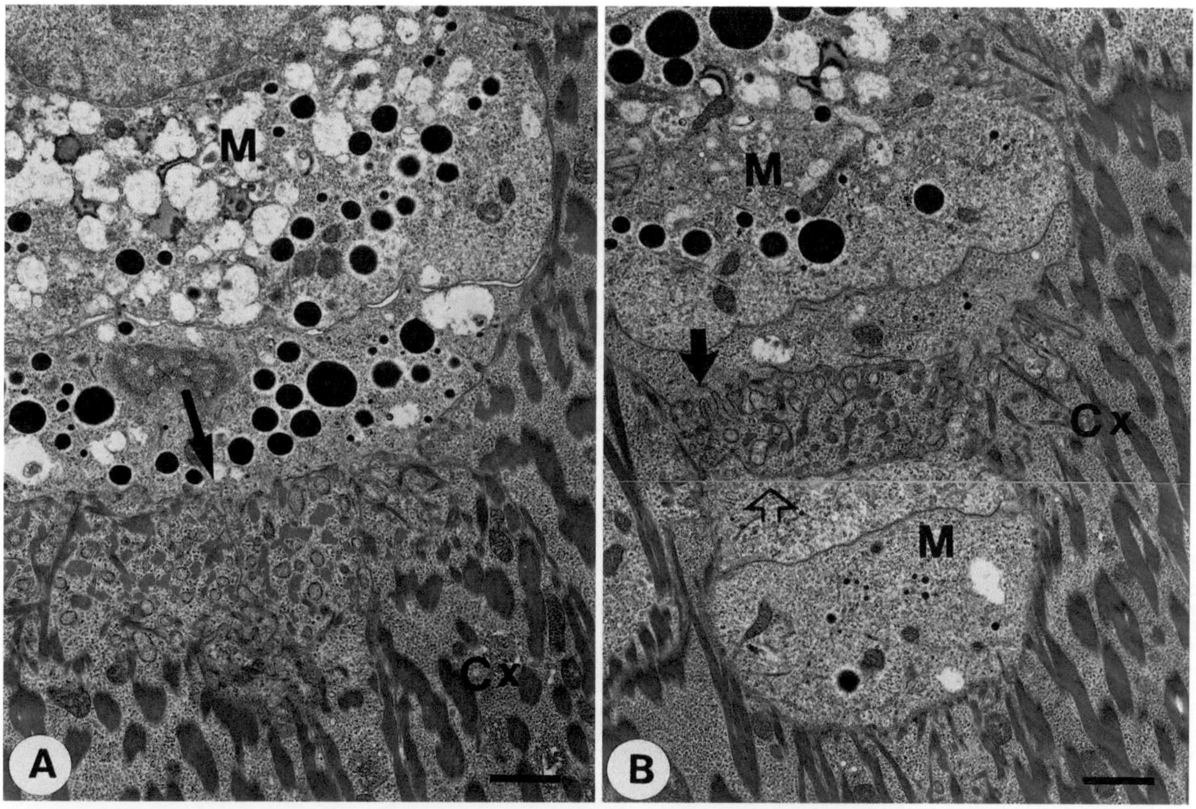

Fig. 2.6. Active areas of medulla cell membranes. The edge and the center of the cells. *M*, medulla; *Cx*, hair cortex. **A** Sagittal section. The *arrow* represents the position of exocytosis-like movement of cell membrane. **B** Longitudinal section. The *black arrow* indicates the active side and the *open arrow* the inactive side of the membrane. Dorsal skin of day 7 rats was used for the experiment. *Bars* 1 μm

Granules and Vesicles

The matured cells of the medulla are filled with many types of granules and vesicles. Arrows in Fig. 2.7A,B designate lysosomes, which contain mitochondria, suggesting that some mitochondria are destroyed by the lysosomal lytic system. As with other tissues of the hair shaft and inner root sheath, mitochondria of the medulla disappear during terminal differentiation (cornification). Although some are shown to be scavenged through the action of lysosomes, others may decompose by themselves without the aid of lysosomes (Fig. 2.7D). In the latter case, the debris may be scavenged by cytoplasmic proteases such as calpains and proteasomes, as well as cathepsins. Lysosomal enzymes and μ-caplain were shown to be colocalized in the area including the medulla during the course of terminal differentiation [17] (see Fig. 1.9C,E,F). They are also colocalized in the keratinizing Henle's layer and keratinizing hair shaft [17], suggesting a common mechanism of terminal differentiation in these tissues.

Various granules that appear in medullary cells are shown in Fig. 2.7C. The vesicle designated by arrow 1 is a multivesicular body that is a kind of

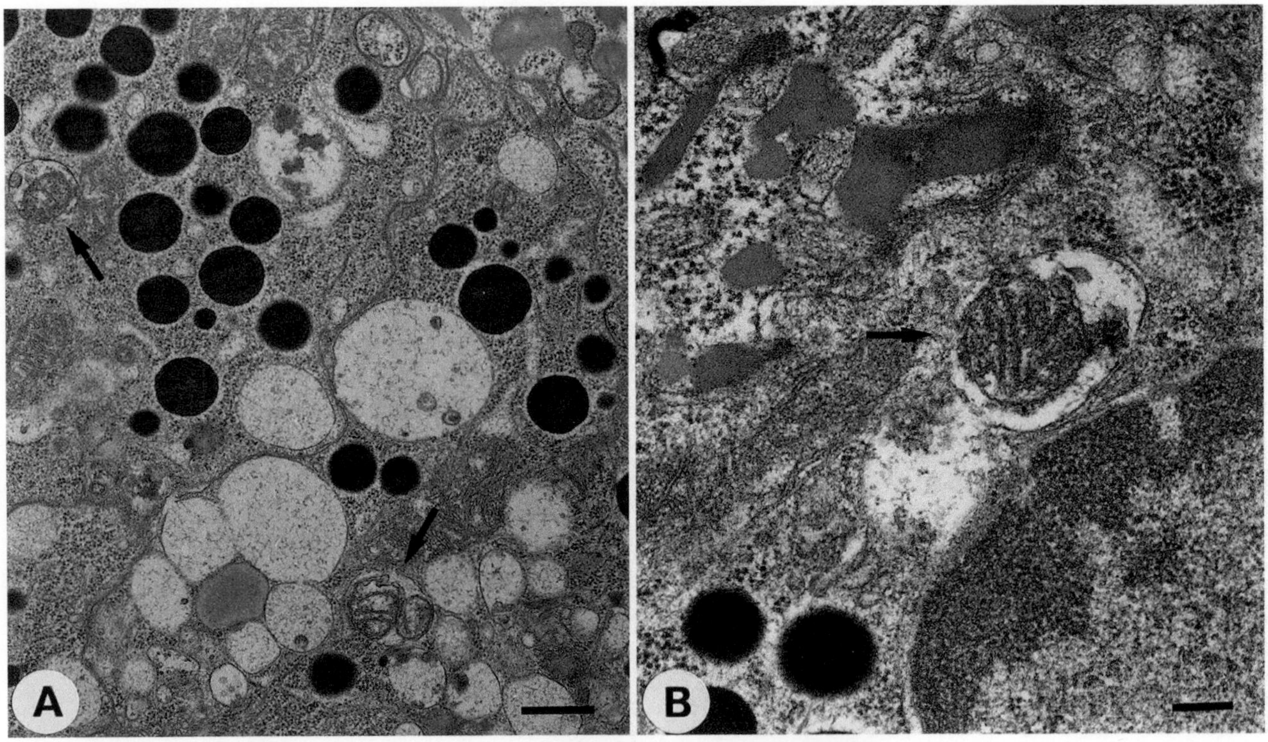

Fig. 2.7. Vesicles and granules in matured cells of medulla. **A** Trichohyalin granules and lysosomes. *Arrows* indicate the inclusion of mitochondria within the lysosomes. **B** Degradation of mitochondria in a lysosome. The position is indicated by the *arrow*. **C** Various vesicles appeared in the cytoplasm of the medulla cell. Explanations for numbered (*1–5*) organelles are given in the text. **D** Self-disintegration of mitochondria in order of *1, 2, 3*. All samples were obtained from dorsal skin of day 14 rats. *Bars* **A, C, D** 500 nm; **B** 200 nm

lysosome. Lysosomes containing inclusions are also shown in Fig. 2.7A,C (number 2). Some of these may have digested almost all the inclusions (Fig. 2.7C, number 3). The vesicle numbered 4 is a major type in the cytoplasm of the mature medulla cell, but neither its history nor function is known. The shape of trichohyalin granules changes as the cell matures. They are approximately spherical in the premature stage (Fig. 2.8A), becoming polymorphic (pleiomorphic) as the cells mature (Fig. 2.8C) and with some associating with others to form doublet or triplet structures (see arrowheads in Fig. 2.8B). Powell and Rogers emphasized that the fusion of trichohyalin granules is one of the most important and characteristic events in the differentiation of the medulla [18]; in my studies, however, fusion was not frequently observed and the number of associations was in any case limited. It seems impossible that the fused mass would fill the cytoplasm of medullary cells. Interestingly, some trichohyalin granules develop a sea urchin-like shape with numerous fine surface protuberances (Fig. 2.8D, arrowheads). I suspect that the granules decompose, followed by molecular reassociation with extensive bonding of proteins to form hard and homogeneous cytoplasmic constructs. The granule highlighted by the arrow in Fig. 2.8D has not been identified.

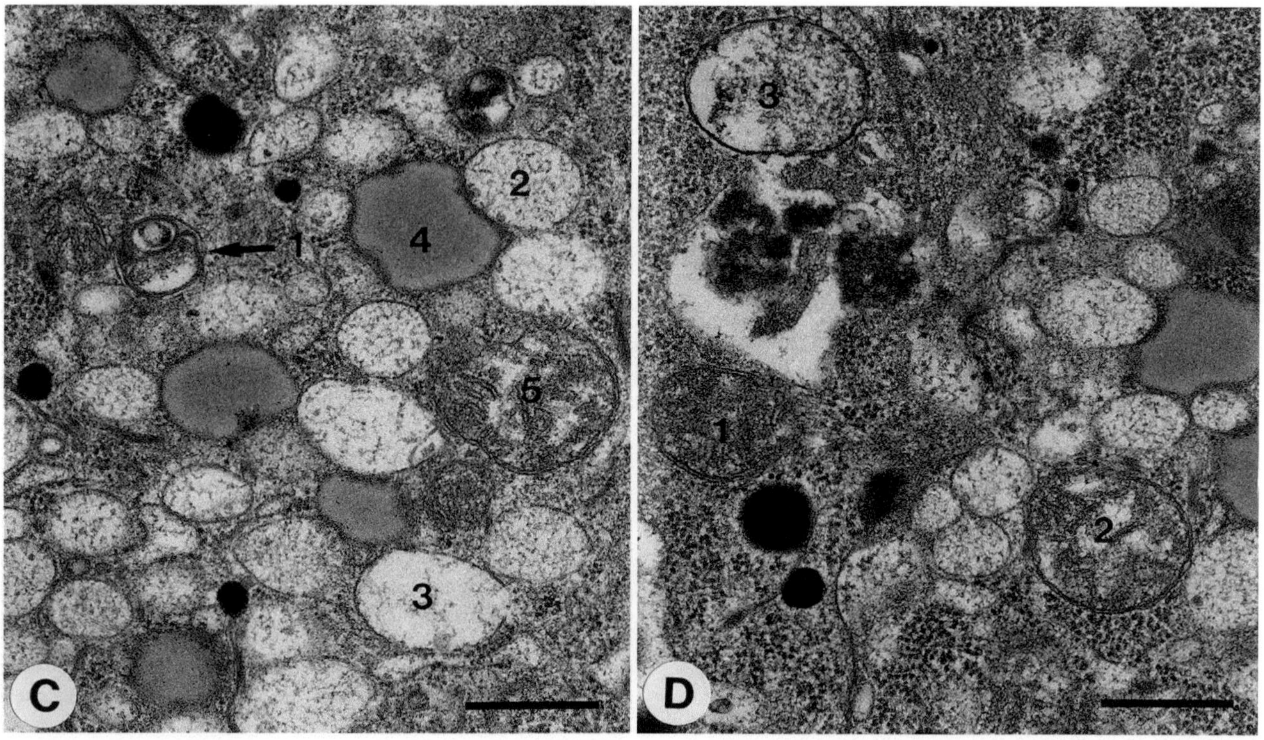

Fig. 2.7.

Groove

Before proceeding to observations on the degeneration process, I briefly linger on the developing medulla to compare its mature cells with younger cells. The cell membrane of the medulla adjacent to the cortex is largely smooth in the premature stages (Fig. 2.9A). When the cells mature, cell-to-cell transfer of the cytoplasm begins as indicated by the arrows in Figs. 2.4A,C and 2.9B. On two-dimensional (2-D) view this process appears as the formation of hollows on both sides of the cell (thick arrows in Fig. 2.9C) and as a groove surrounding the cell on 3-D. After the transfer subsides, the groove is filled with tonofilament bundles

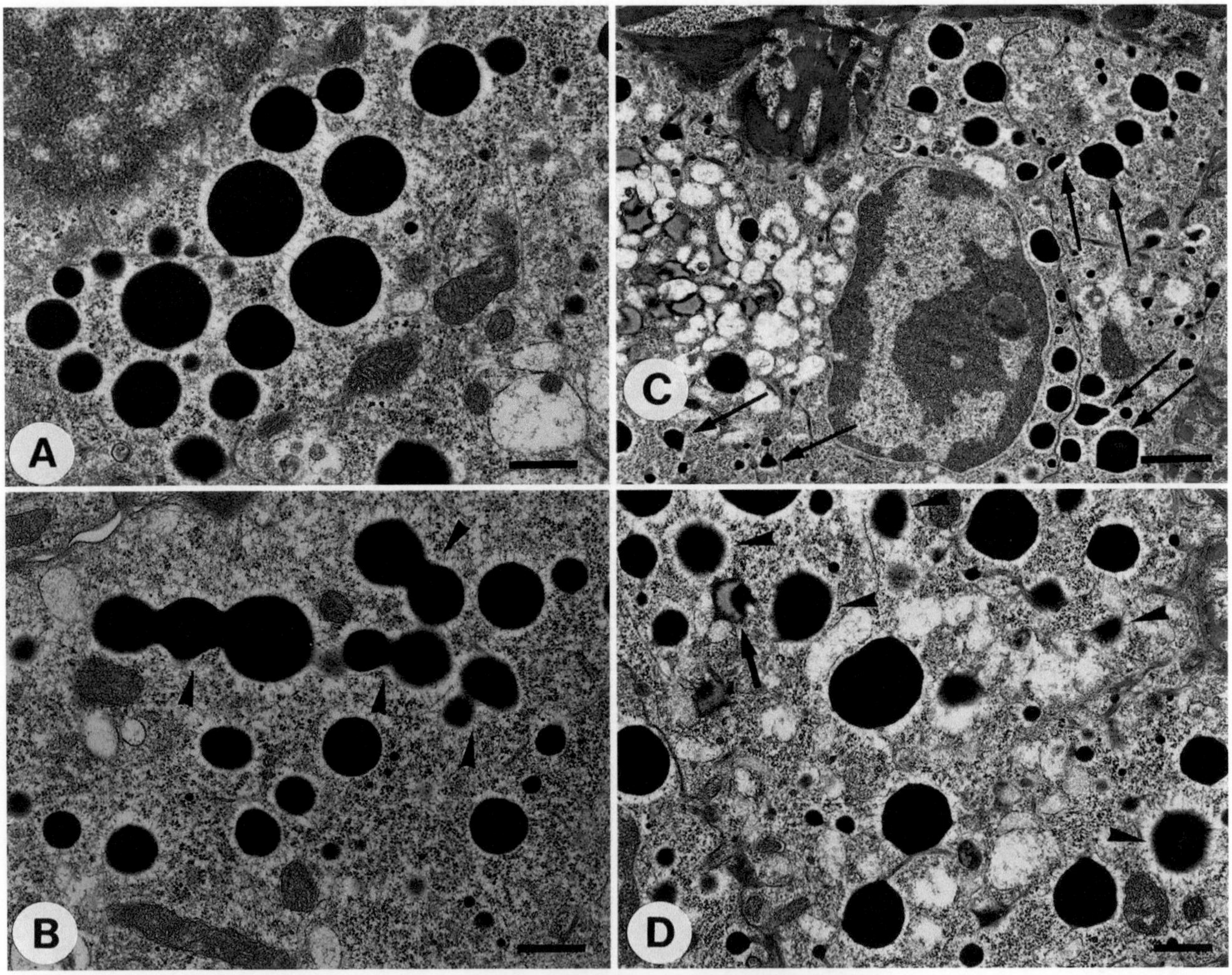

Fig. 2.8. Transformation and association of trichohyalin granules in matured cells of medulla. **A** Normal spherical granules. **B** Association of granules (see *arrowheads*). **C** *Arrows* show that the shape of the granules became not spherical. **D** Surface of some granules designated by *arrowheads* became rugged. Granule highlighted by *arrow* was not identified. Samples were obtained from dorsal skin of day 7 rats. *Bars* **A, B, D** 500 nm; **C** 1 μm

of cortex cells, as indicated by the thin arrows in Fig. 2.9C. An enlarged view is provided in Fig. 2.10A in which the bundles are indicated by thick arrows. It is notable that that the orientation of the bundles in the groove is different from that of other bundles in the cortex: in contrast to the latter, aligned according to the longitudinal axis of the cell, the former radiates and appears as shoring to stabilize the groove.

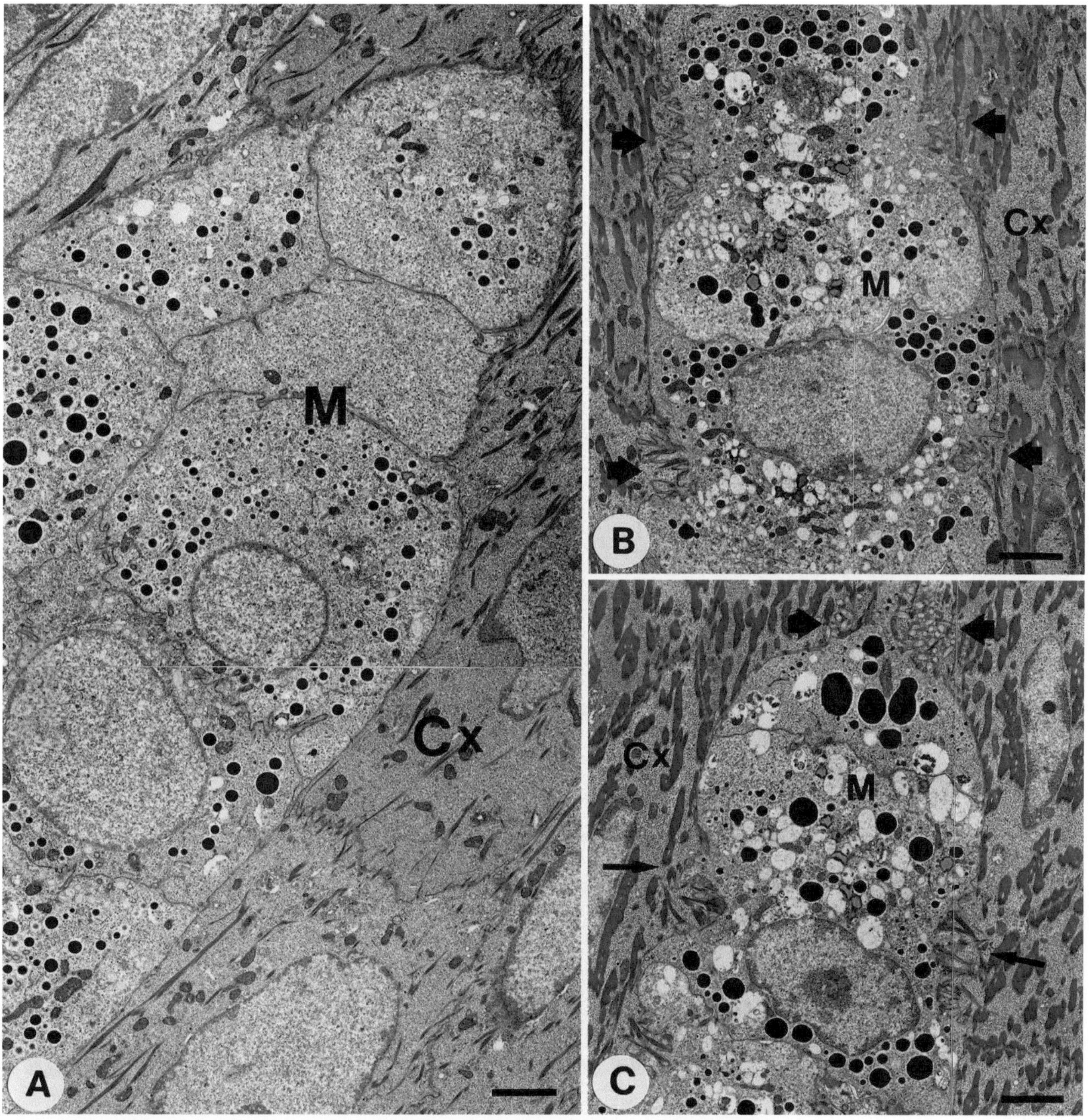

Fig. 2.9. Development of the groove in medulla cell. *M*, medulla; *Cx*, hair cortex. **A** Young medulla cells as controls. **B** *Thick arrows* designate the ruffling of cell membrane that seems to be an initial event to make a groove. **C** *Thick arrows* designate active areas of the membrane. Grooves become deep. *Thin arrows* show the formation of tonofilament bundles in the grooves. Samples were obtained from dorsal skin of day 7 rats. *Bars* 2 μm

Although anomalous active movement of membranes between medulla and cortex occurs, this movement never breaks the adhesion between them. In contrast, medulla cells became detached from adjacent medulla cells (Fig. 2.10B). Some investigators term the resulting intercellular spaces fusi [19, 20]. The positions of detachment are indicated by arrows in Fig. 2.10B. The boundary of cells encircled by the ellipse in Fig. 2.10B shows an intricate structure which develops just prior to detachment of the membranes. This special movement of cell surfaces ceases after separation of the cell membranes is accomplished. It thus seems that the physiological role of the medulla cell membrane differs between its cortex and medulla sides. Even after the apparent death of medulla cells, adherence of the cortex side membrane to the cortex cells seems to persist, in contrast to the wide margins that appear between medulla cells (see Fig. 2.12).

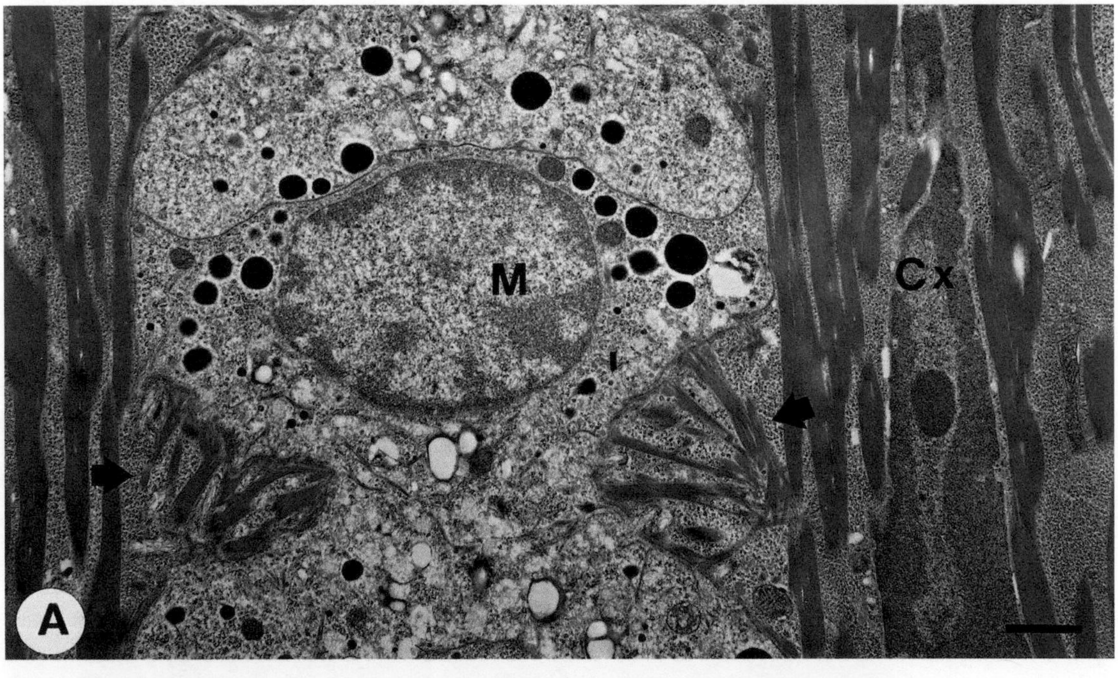

Fig. 2.10. Structural changes of medulla accompanied by its maturation or programmed death. Longitudinal view of a dorsal skin hair obtained from a day 7 rat. *M*, medulla; *Cx*, hair cortex. **A** Morphology and orientation of tonofilament bundles in the hollows of medulla cells. **B** Separation of medulla cells. *Encircled area* seems to be in the process of the separation. *Bars* 1 μm

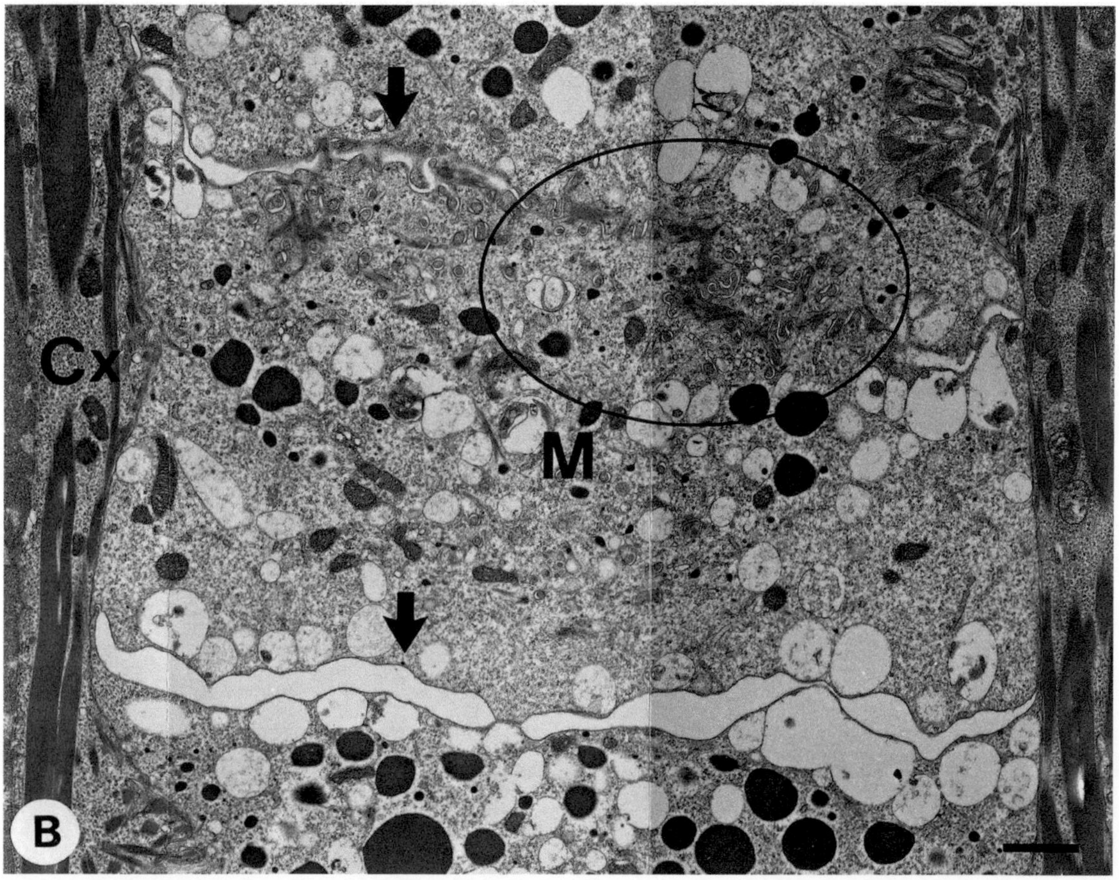

Fig. 2.10.

Programmed Cell Death

The notion of apoptosis cannot be ignored when cell death is discussed. The terminal differentiation of the medulla is somewhat different from that of other hair shaft cells. In place of bundles of keratin fibers, the cytoplasm of the medulla is filled with proteins intercalated with isopeptide bonds, as described earlier (the section on Trichohyalin in this chapter). Morphological changes in the nucleus in the early stages are similar to those in the original example of apoptosis presented by Wyllie et al. [21]. The chromatin seems to aggregate, and the outline of the nucleus is abnormally convoluted (Fig. 2.11).

The process thereafter deviates from that of typical apoptosis. The chromatin swells and diffuses evenly in the nucleoplasm in later stages, resulting in a structure-less nucleus (Fig. 2.12). Neither crescent shaping of chromatin nor fragmentation of chromatin or of the nucleus is observed. According to the original study, apoptosis occurs sporadically [20]; here, in contrast, the pattern of the death of medulla cells occurs contiguously in the central tract of the hair. In view of its morphological features, I consider the death of these cells to resemble that of embryonic erythroid cells [22] rather than typical apoptotic cell death.

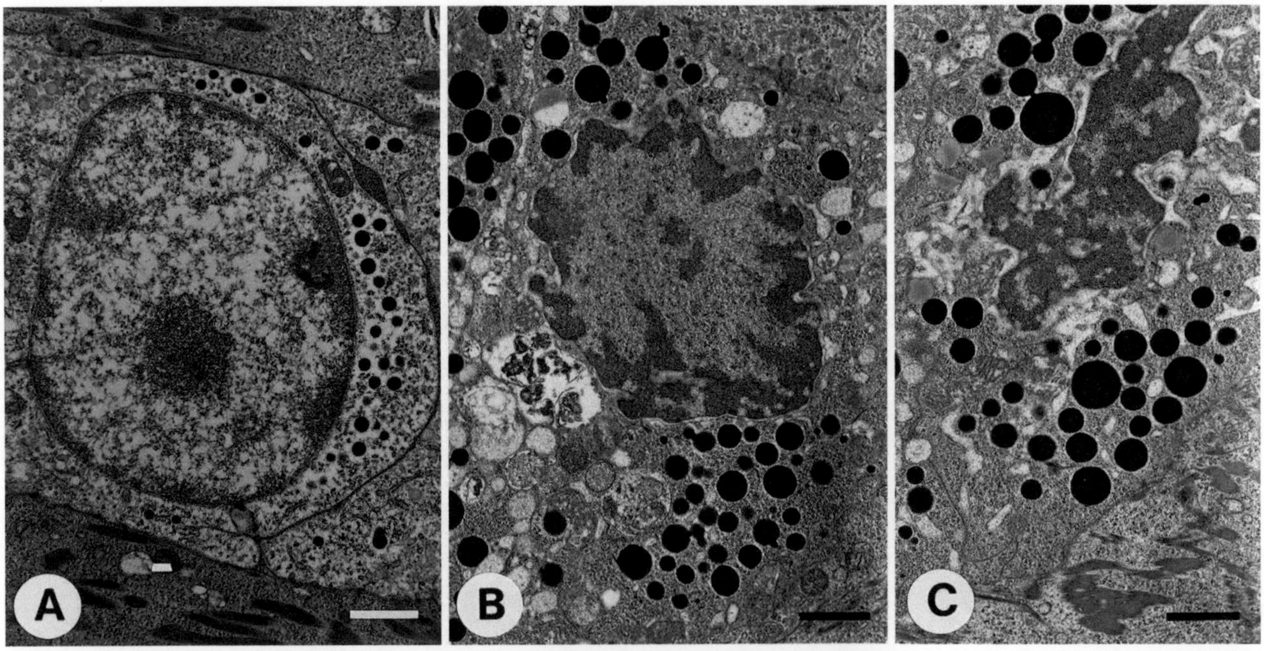

Fig. 2.11. Morphological change of nucleus during the maturation of medulla. **A** Nucleus of a premature cell from dorsal skin of a day 7 rat. **B, C** Nuclei of mature cells from the dorsal skin of day 14 rats. *Bars* 1 μm

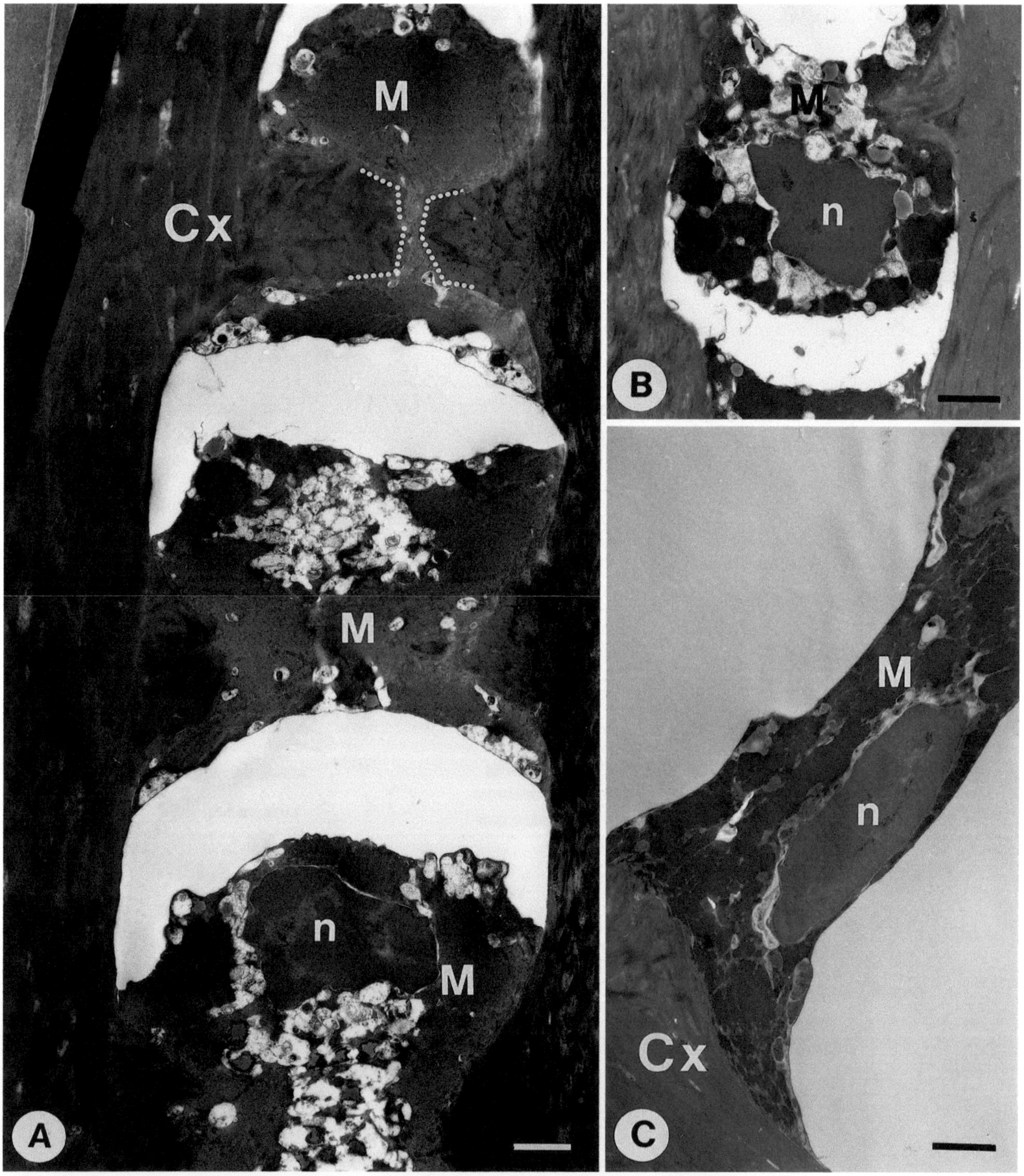

Fig. 2.12. Programmed death of medulla. *M*, medulla; *Cx*, hair cortex; *n*, nucleus. *Dotted lines* designate the boundary between M and Cx. Degeneration of medulla proceeds from **A** to **B** to **C**, which retains very thin cell remnants. Samples were obtained from the dorsal skin of day 14 rats. *Bars* 1 μm

The terminal differentiation of medulla cells is schematically summarized in Fig. 2.13. The extruded hard keratin bundles form partitions resembling the knots of a bamboo tree (asterisks in Fig. 2.13B). These knots reinforce the medullary pipe structure. This special structure thus provides both strength and an insulative cavity to the hair.

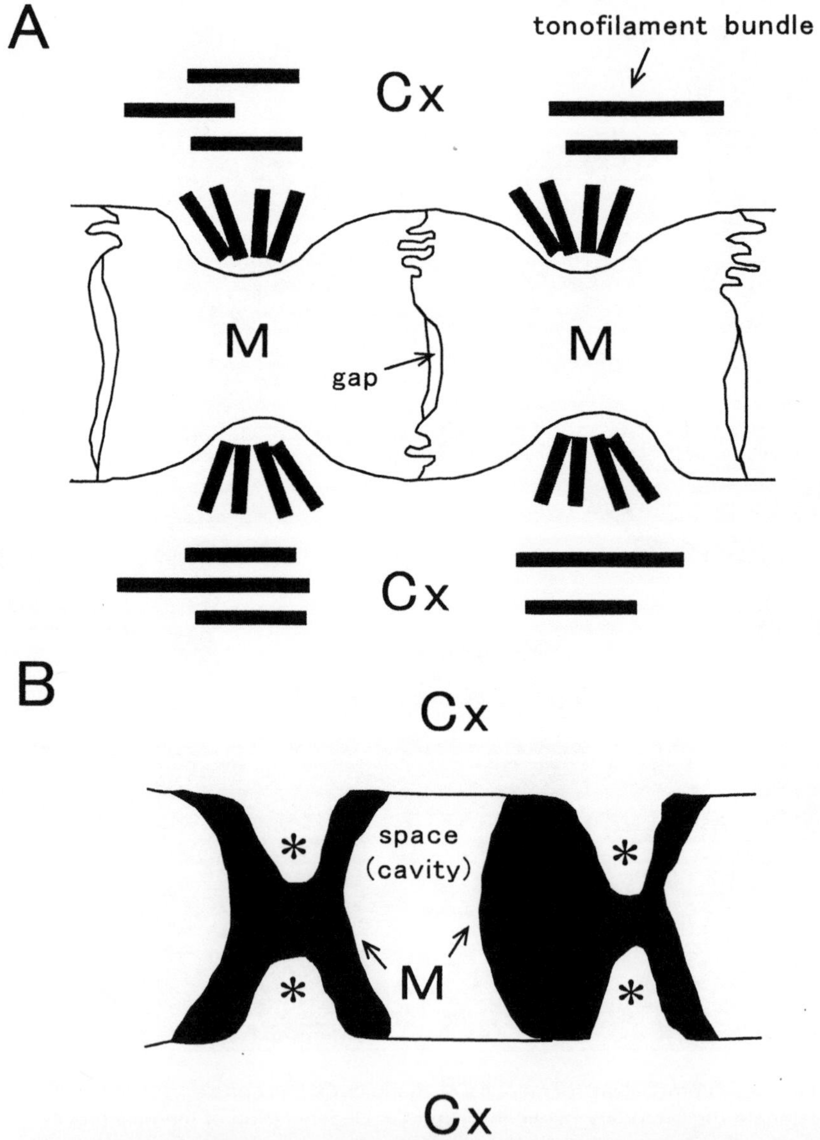

Fig. 2.13. Schema to make cavities in the cornifying process of medulla cells. *M*, medulla; *Cx*, hair cortex; *asterisks*, extrusion of hair cortex

References

1. Bradbury JH, Leeder JD (1970) Keratin fibres. IV. Structures of the cuticle. Aust J Biol Sci 23:843–854
2. Bacha WJ, Wood LM (1990) Color atlas of veterinary histology. Lea & Febiger, Philadelphia
3. Hisa: http://www.polarbearc.com/kouza/kouza_6.html (1997–2003) in Japanese. e-mail address: info@polarbear.com
4. Maeda H (199) Inspection of hair. In: Takatori T (ed) Essentials of forensic medicine, 3rd edn (in Japanese). Ishiyaku Shuppan, Tokyo, pp 299–301
5. Forensic science and the legal process. Lecture: Trace and contact evidence. cntd. http://www.cpes.sussex.ac.uk/fslp/lnotes/lecture7.pdf, autumn 2002
6. Halal J (2002) Structure of hair. In: Hair structure and chemistry simplified, 4th edn. Miladay, New York, p 60
7. Deedrick DW (2000) Hairs, fibers, crime, and evidence. Part I. Hair evidence. In: Forensic science communications, vol 2, no 3. http://www.cpes.sussex.ac.uk/fslp/lnotes/lecture7.pdf
8. Fawcett DW (1986) A textbook of histology. Saunders, Philadelphia, pp 563–565
9. Bacha WJ Jr, Wood LM (1990) Color atlas of veterinary histology (Japanese version). Lea & Febiger, Philadelphia, pp 88–90
10. Harding HWJ, Rogers GE (1971) ε-(γ-Glutamyl) lysine cross-linkage in citrulline protein fractions from hair. Biochemistry 10:624–630
11. Rogers GE, Kuczek ES, Mackinnon PJ, Presland RB, Fietz MJ (1989) Special biochemical features of the hair follicle. In: Rogers GE, Reis PJ, Ward KA, Marshall RC (eds) The biology of wool and hair. Chapman & Hall, London, pp 69–85
12. Fietz MJ, McLaughlan CJ, Campbell MT, Rogers GE (1993) Analysis of the sheep trichohyalin gene: potential structural and calcium-binding roles of trichohyalin in the hair follicle. J Cell Biol 121:855–865
13. Lee SC, Kim IG, Marekov LN, O'Keefe EJ, Parry DAD, Steinert PM (1993) The structure of human trichohyalin. Potential multiple roles as a functional EF-hand-like calcium-binding protein, a cornified cell envelope precursor, and an intermediate filament-associated (cross-linking) protein. J Biol Chem 268:12164–12176
14. Rhodin JAG (1974) Histology. A text and atlas. Oxford University Press, New York
15. Ito M, Hashimoto K Trichohyalin granules in hair cortex. J Invest Dermatol 79:392–398
16. Martin-Verdeaux S, Pombo I, Iannascoli B, Roa M, Varin-Blank N, Rivera J, Blank U (2003) Evidence of a role for Munc 18-2 and microtubules in mast cell granule exocytosis. J Cell Sci 116:325–334
17. Morioka K, Sato-Kusubata K, Kawashima S, Ueno T, Kominami E, Sakuraba H, Ihara S (2001) Localization of cathepsins B, D, L, LAMP-1 and μ-calpain in developing hair follicles. Acta Histochem Cytochem 34:337–347
18. Powell BC, Rogers GE (1997) The role of keratin proteins and their genes in the growth, structure and properties of hair. In: Jolles P, Zahn H, Hoker H (eds) Formation and structure of human hair. Birkhauser, Basel, pp 59–148
19. Hausman LA (1932) Am Nat 46:461
20. Roth SI, Helwig EB (1964) The cytology of the cuticle of the cortex, the cortex, and the medulla of the mouse hair. J Ultrastruct Res 11:52–67
21. Wyllie AH, Kerr JFR, Currie AR (1980) Cell death: the significance of apoptosis. Int Rev Cytol 68:251–306
22. Morioka K, Tone S, Mukaida M, Takano-Ohmuro H (1998) The apoptotic and nonapoptotic nature of the terminal differentiation of erythroid cells. Exp Cell Res 240:206–217

Hair Cortex and Hair Cuticle

3

Initial Signs of Keratinization

The first morphological sign of the keratinization of the hair cortex (Cx) is the formation of small bundles of fibers in the cell. These often appear near the periphery of the cell (arrowheads in Fig. 3.1), but can also be found everywhere in the cytoplasm, as seen in the lower part of the same Cx cell in Fig. 3.1. The medulla is not seen in this section; it is difficult to determine whether this follicle has no medulla or whether the section is simply higher than the top of the medulla. In Fig. 3.1, one Cx cell seems to be embraced by three hair cuticle (Ce) cells. These Ce cells exhibit no signs of cornification. Desmosomes are observed extensively in the early follicle (Fig. 3.1; an enlarged view of another example is shown later in Fig. 4.13B), suggesting that they are required to maintain cellular association, particularly in the early stages. Generally, the Cx starts its differentiation earlier than the Ce, at least from the morphological point of view. Some (but not all) cells of the cuticle of the internal root sheath (Ci) in Fig. 3.1 are just starting to undergo cornification through the production of tonofilaments in their cytoplasm. The boundary between the Ce and Ci, in other words the boundary of the hair shaft and internal root sheath, is designated by the dotted line in Fig. 3.1. Five Ci cells seem to encircle the Ce. All cells of the He and Hx layers seem to possess both trichohyalin granules and tonofilaments. These observations suggest that the Ce starts to cornify later than the inner root sheath cells. The boundaries between the hair shaft (HS) and inner root sheath (IRS) and between the inner root sheath and outer root sheath (ORS) are largely smooth and exhibit no sharp or deep bends, distinguishing them from other cellular boundaries, which are generally complicated. The simplified image in Fig. 3.2 reveals that the HS–IRS and IRS–ORS boundaries are distinguishable in a geometric way, independently of any consideration of the features of cellular differentiation.

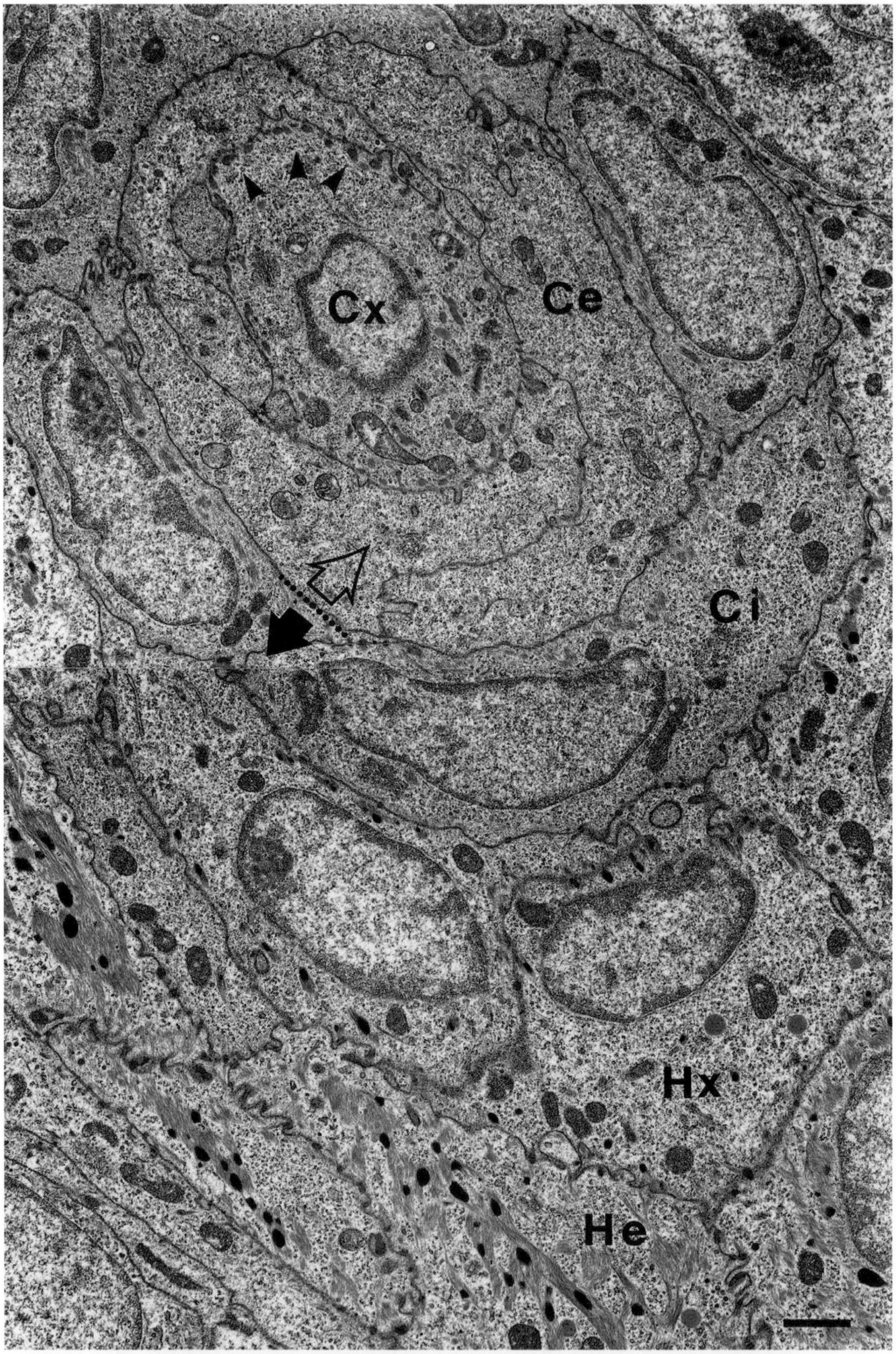

Fig. 3.1. Initial sign of keratinization in hair cortex and surrounding immature cells. *Cx*, hair cortex; *Ce*, hair cuticle; *Ci*, cuticle of inner root sheath; *Hx*, Huxley's layer; *He*, Henle's layer; *open arrow*, hair shaft; *black arrow*, internal root sheath. *Dotted line* designates the boundary between hair shaft and internal root sheath; *arrowheads* indicate small bundles of fibers near the cell boundary. Samples were obtained from the face skin of day 2 rats. *Bar* 1 μm

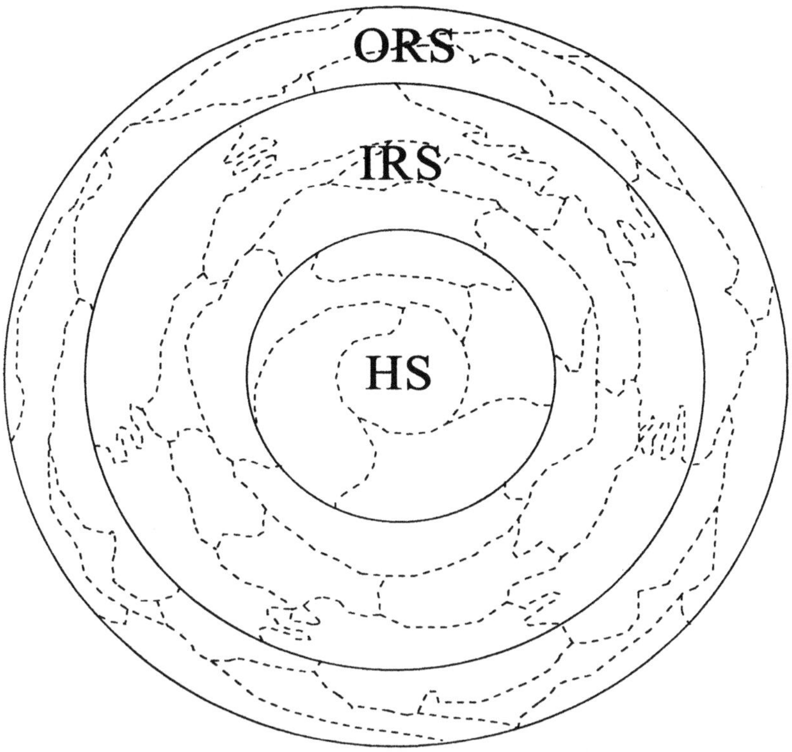

Fig. 3.2. Geometrical identification of outer root sheath, inner root sheath, and hair shaft in the immature hair follicle presented schematically. *ORS*, outer root sheath (with companion layer in this case); *IRS*, inner root sheath; *HS*, hair shaft. *Solid lines* represent the boundaries of ORS, IRS, or HS; *dotted lines* represent the boundaries of cells

Accumulation of Keratin Fibers

The Ce cells start to produce small granules in their cytoplasm after the Cx cells have accumulated many tonofilament bundles (Fig. 3.3). This granule production is the first morphological sign of the differentiation of the Ce. These small granules are not the trichohyalin granules discussed next, but probably consist mainly of special keratins [1, 2]. They do not disappear during the differentiation of Ce cells, but enlarge and migrate to the outer region of the cells [3]. Details of the keratinization of Ce are described later in this chapter. At the time and height of Fig. 3.3, Henle's layer (He) has been keratinized densely. Huxley's layer (Hx) accumulates trichohyalin granules of a variety of sizes, whereas the cuticle of the internal root sheath (Ci) accumulates small granules only.

Identification of the tissues becomes easier as the differentiation proceeds to the stages represented by Fig. 3.3. The Cx in this figure is surrounded by three or four layers of Ce, which are themselves surrounded by the internal root sheath layers. In this section, each sublayer of the internal root sheath, namely He, Hx, and Ci, consists of a cellular monolayer (Fig. 3.3). The chromatins in the nuclei (designated as n) of the hair cortex cells appear to diffuse slightly. Given that Ce did not respond to antitrichohyalin antibody in the immunohistochemical study of O'Guin et al. [2], the granules accumulated in the Ce are suggested to be not trichohyalin granules. On indirect immunofluorescence staining of human hair follicles, the antitrichohyalin monoclonal antibodies responded to internal root sheath layers and medulla, but not to Ce [2].

In contrast to epidermal keratinocytes in the spinous or granular layers that have a special perinuclear compartment in which granules and fibers are scarcely found, every keratinizing hair cell exhibits normal morphology in its perinuclear area (e.g., compare the Cx cells in Fig. 3.3 with the keratinocytes in Reference 4). Some nuclei in the Cx are polymorphic as early as this stage of keratinization.

Fig. 3.3. Accumulation of keratin fiber bundles in hair cortex and status of surrounding tissues. *Cx*, hair cortex; *Ce*, hair cuticle; *Ci*, cuticle of inner root sheath; *Hx*, Huxley's layer; *He*, Henle's layer; *n*, nucleus. Samples were obtained from the face skin of day 2 rats. *Bar* 1 μm

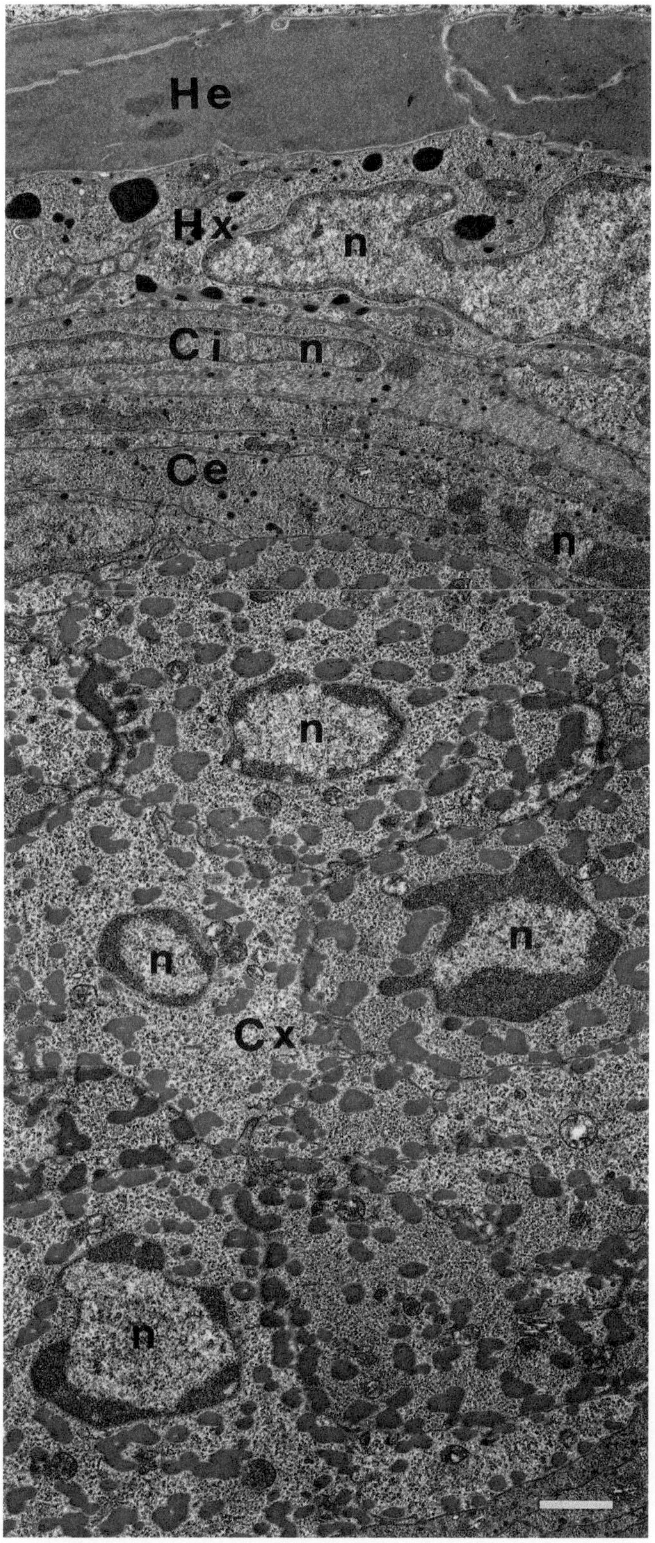

Fig. 3.3.

Keratin Filament Bundles

The formation of keratin bundle is one of the key processes in the hair growth, and its mechanism has been the center of interest for hair investigators. Besides biochemical or molecular biological study, electron microscopy may also contribute to elucidate the mechanism. Dense areas clearly distinguishable from normal fibrous areas are often observed during the growth of keratin filament bundles in the Cx (arrows in Fig. 3.4C). These areas are widely distributed in the cortex cells (Fig. 3.4A). One explanation is that they are the remnants of lysosomal digestion, and lysosome-like vesicles inside the bundles are in fact seen in Fig. 3.4B. Another possibility is that they are precursors of the components of the bundles or are the prekeratination areas accumulating precursor components. It is interesting that these areas are usually observed around the center part of the bundles (Fig. 3.4A,C).

In contrast to the active intrabundle autophagy in Cx cells, autophagy of the mitochondria is not frequently observed in those cells. This finding may suggest that mitochondria cause their own disintegration (natural degradation) via activation of a lipoxygenase [5] during the differentiation of Cx cells. This degeneration is likely coordinated with cytoplasmic proteases. Of course, one can also find fully fibrous bundles having largely uniform structures (Fig. 3.4D). In addition to the Cx, the Ce is also shown to be filled with lysosome-like vesicles (Fig. 3.4B). Setting aside the question of the fate of mitochondria, which is discussed later in this chapter, I believe that the lysosomal system in some cell types plays an important role in the "cytoplasmic conversion" or terminal differentiation that involves the overall replacement of the cytoplasmic protein species [6].

Fig. 3.4. Structure of keratin bundles in hair cortex cells. *Cx*, hair cortex; *Ce*, hair cuticle; *Ci*, cuticle of inner root sheath;. *Hx*, Huxley's layer. **A, C** Heterogeneous staining pattern of the fiber bundles in the cortex cell. **B** Vesicles or lysosomes in the fiber bundles. **D** Homogeneous staining pattern of the fiber bundles. Regrettably, parts **B** and **D** show knife marks. Samples were obtained from the dorsal skin of day 7 (**A, C**) or day 14 (**B, D**) rats. *Bars* **A, B** 1 μm; **C, D** 500 nm

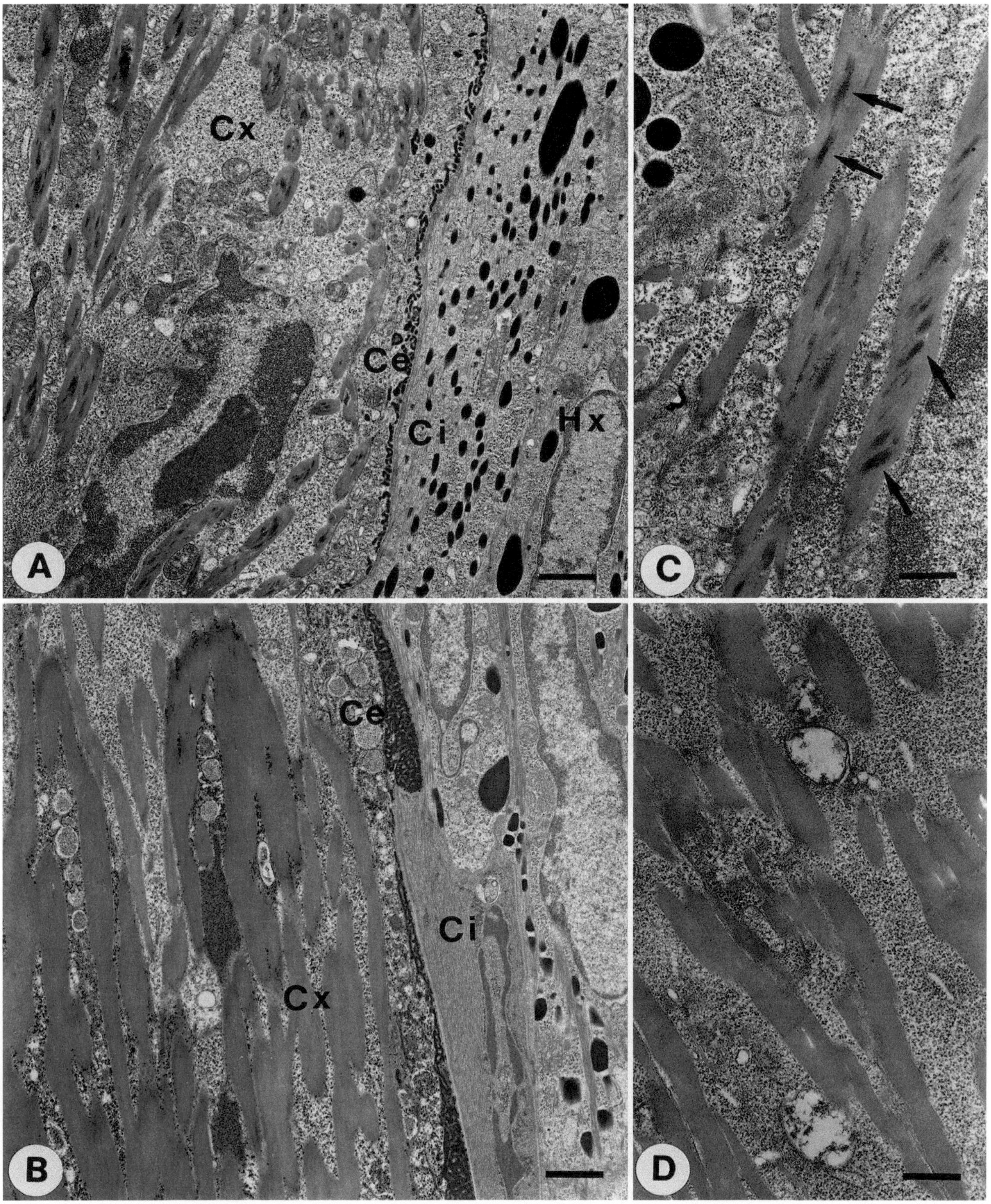

Fig. 3.4.

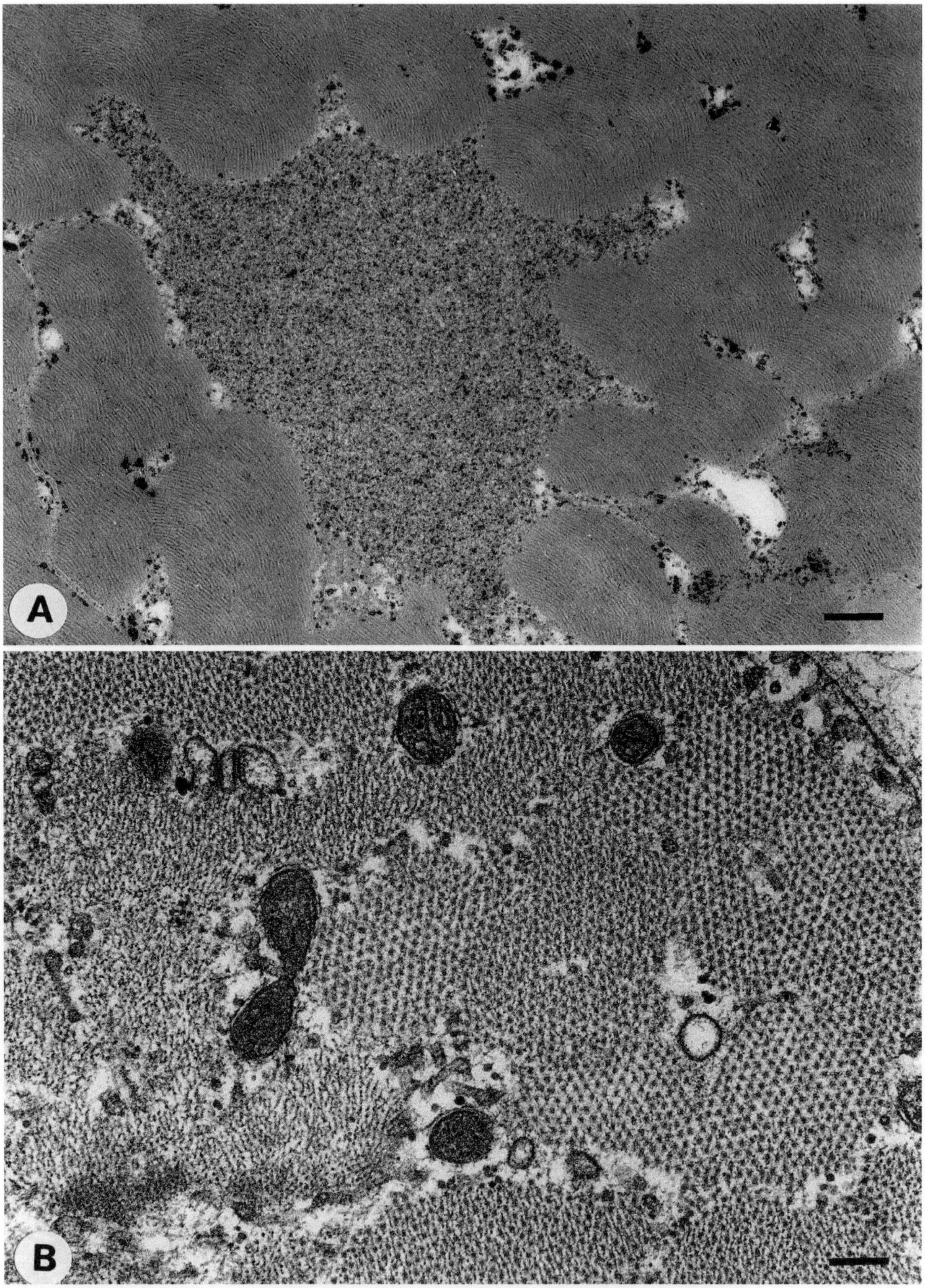

Fig. 3.5. Fine structures of the bundles. **A** Fingerprint-like staining pattern characteristic of keratin filament in the cortex cell. **B** Cross section of keratin fibers in the cortex cell. Samples were obtained from the dorsal skin (**A**) or facial skin (**B**) of day 2 rats. *Bars* 200 nm

The structure of the bundles is detailed in Fig. 3.5. Fingerprint-like structures characteristic of bundles of hard keratin filaments [7] are observed (Fig. 3.5A). In addition to the fingerprint structure, I have observed a hexagonal arrangement of unique components (Fig. 3.5B) in place of the cylindrical lattice in wool fiber [7, 8] and in adult rat hair [9]. This arrangement probably results from the immaturity of the structure of the bundles. Each of the structural units in Fig. 3.5B consists of a core and spines. The arrangement of the cores is schematically shown in Fig. 3.6; in this figure, each unit is drawn for simplicity as an association of 8 fibers, but the actual number of associated fibers varies and is generally about 20 or more. Patterns A and B in Fig. 3.6 are assumed to be a basic structure and a more mature one, respectively. Both patterns are seen in Fig. 3.5B. Interspaces are expected to be gradually filled with the spines (associated proteins) as part of the progress of cornification. The major biochemical components of these bundles are hard keratins and keratin-associated proteins (KAP) [8], but the process of their formation has not been fully elucidated.

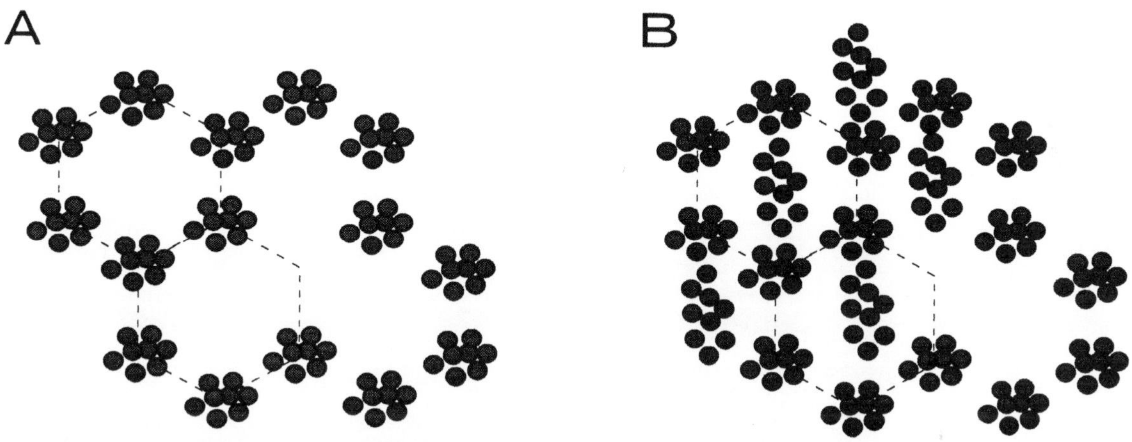

Fig. 3.6. Hexagonal arrangement of fibers and their maturation. Although 8 fibers are shown tentatively associated in this figure for simplification, the real number of associated fibers is estimated to be 20 or more (diverse) based on the analysis of the enlarged view of Fig. 3.5B. Patterns of **A** and **B** are assumed to be a basic structure and a mature one, respectively. Both patterns are seen in Fig. 3.5B

Morphological Change of the Nucleus

Figure 3.7 shows the morphological changes of nuclei in the hair cortex (A–C) and hair cuticle (D–G). Normal morphology of the nucleus and chromatin with a distinct nucleolus is observed when the cortex cells are at an immature stage, that is, when tonofilaments are still thin and sparse (Fig 3.7A). The chromatin assumes a pasty appearance and the contour of the nucleus becomes amorphous when accumulation of thick bundles of tonofilaments is seen (Fig. 3.7B), after which the nuclei appear structure less and are evenly stained (Fig. 3.7C). This pattern is similar to that in the medulla. Structure-less cortex nuclei have also appeared in a textbook [9]. The morphological changes in nuclei of the hair cuticle are essentially similar to those of the medulla and cortex (Fig. 3.7D–G). Because cuticle cells appear as thin disks, in contrast to the spindle shape of the cortex cells, the nuclei of the former are also thin. The chromatin of the hair cuticle cell often shows a characteristic zebra pattern (Fig. 3.7F) during the transitional period of differentiation. The chromatin then diffuses evenly in the nucleoplasm to form a structure-less nucleus (Fig. 3.7G). Nuclei of the cortex and cuticle cells in the hair shaft thus display a similar process of morphological change, involving the disintegration of the chromosomal structures.

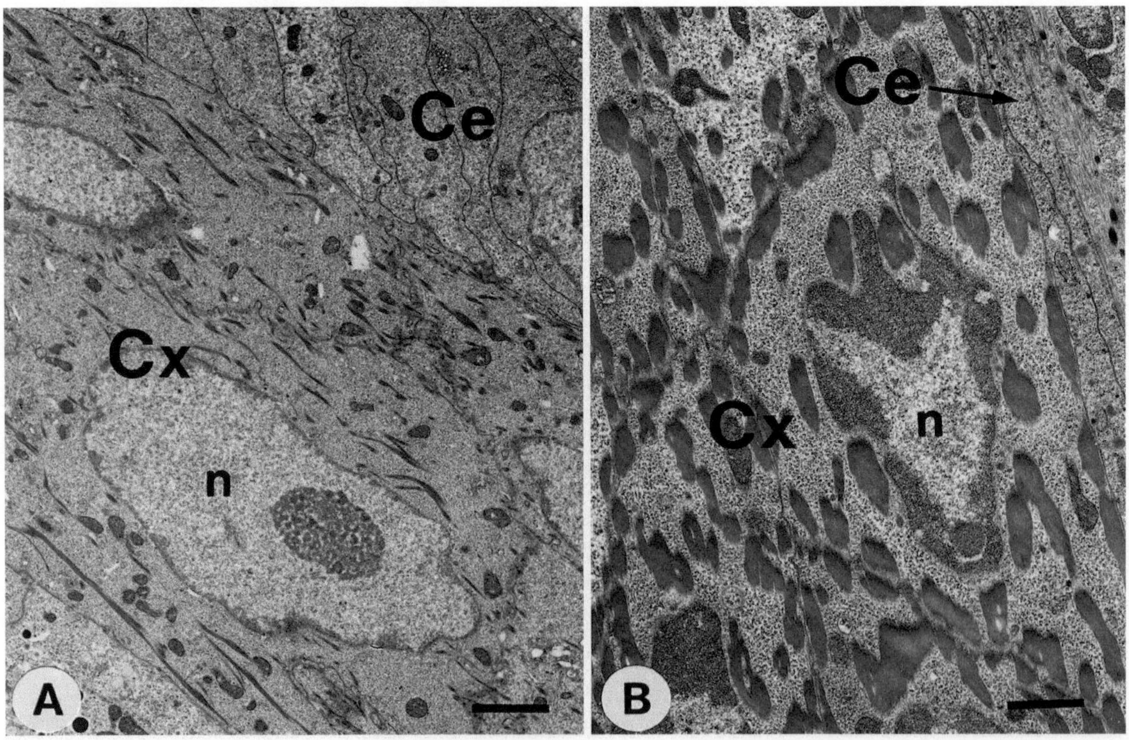

Fig. 3.7.

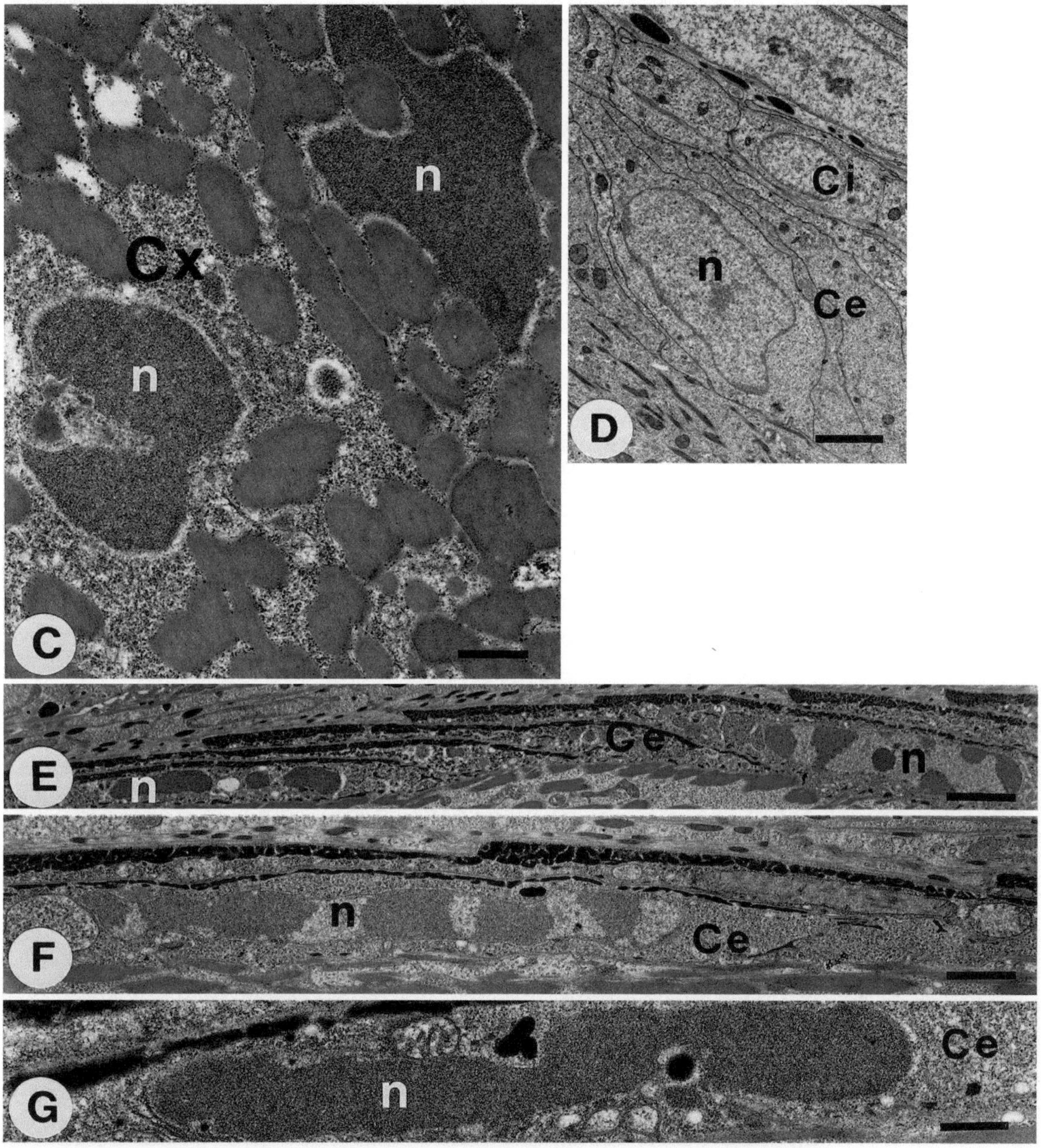

Fig. 3.7. Morphological change of nucleus in hair cortex and cuticle. *Cx*, hair cortex; *Ce*, hair cuticle; *n*, nucleus; *Ci*, cuticle of inner root sheath. **A** A cortex cell in the early stage of the keratinization. **B** Thick bundles of tonofilaments are formed in a cortex cell. **C** The cytoplasm of a cortex cell is about to be filled with tonofilament bundles. **D** A cuticle cell before keratinization. **E, F, G** Progress of the keratinization of cuticle cells and concomitant morphological change of their nuclei. Samples were obtained from the dorsal skin of day 7 rats except **C**, which was from face skin of a day 2 rat. *Bars* **A** 2 μm; **B** 1 μm; **C** 0.5 μm; **D, E** 2 μm; **F** 1 μm; **G** 0.5 μm

Mitochondria

All mitochondria in keratinizing tissues are destined to degenerate during the process of terminal differentiation. Many disintegrating mitochondria are seen in the cortical and cuticular cells in Fig. 3.8A. Traces of lamellar structures are detectable in these mitochondria, as designated by the arrows in Fig. 3.8B. They seem to have lost their double membrane structure, which is replaced instead by a sharply contrasting single membrane (see Figs. 3.4D, 3.8A,B) [6]. The biochemical background of this morphological change, particularly the role of the lysosomal system (if any), remains an unsolved problem. In any case, the loss of mitochondria results in the cessation of major informational flow in the apoptotic cascades, because Bcl-2 family proteins, cytochrome *c*, and some other mitochondrial proteins are known to control apoptosis [10–13]. The other cytoplasmic components of the cortical and cuticular cells, including endoplasmic reticulum, may be degraded within the vesicles or multivesicular bodies designated by the arrowheads in Fig. 3.8B,C,D [see also Reference 13]. These operations of terminal differentiation, which I call cytoplasmic conversion, provide amino acids and other nutritional components for the production of keratin fibers or other constituents of terminally differentiating cortical and cuticular cells. By this I mean that the scavenging of organelles is not only a degeneration process but also implies an efficient new process in reconstitution of the cytoplasm for keratinization in skin and hair, hemoglobinization in erythrocytes, crystallinization in eye lens, and so on [4, 6, 14, 15].

Before the disappearance of the mitochondria, it is possible that agents of the apoptotic machinery such as caspase family enzymes operate automatically to transiently deform the nucleus and chromatin. As this machinery operates for only a limited time, I do not consider that its effect is predominant. Some investigators have claimed that the marooned apoptosis-promoting factors may act automatically, but such an uncontrolled operation would result in uncontrolled, inappropriate, and prompt cell death. On this basis, significant participation of the apoptotic system in the final process of cornification of the hair follicle seems unlikely. Note that this idea does not include the degeneration of the hair follicle in the hair cycle (catagen), which is a different and special process.

Fig. 3.8. Degeneration of mitochondria and conversion of cytoplasmic components. *Cx*, hair cortex; *Ce*, hair cuticle; *M*, medulla. **A, B** Degeneration of mitochondria (see *arrows* in **B**). The *arrowhead* in **B** denotes a multivesicular body (lysosome). **C** Vesicles (*arrowheads*) in hair cortex and hair cuticle. **D** Vesicles (*arrowheads*) in hair cortex. Samples in **A**, **B**, and **D** were obtained from the dorsal skin of day 7 rats; that in **C** was from the dorsal skin of day 14 rats. *Bars* **A, C** 1 μm; **B, D** 500 nm

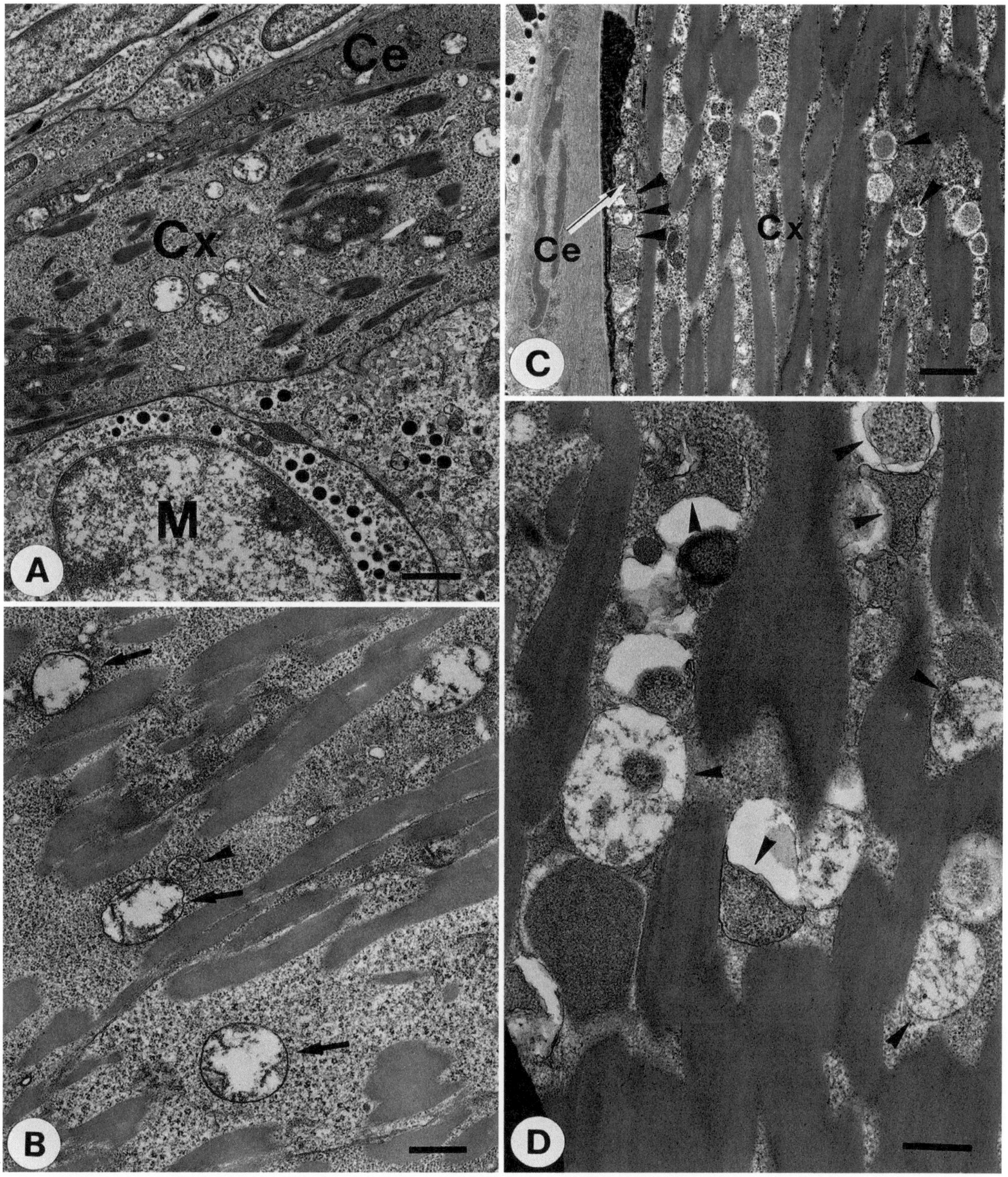

Fig. 3.8.

Keratinization of the Hair Cuticle

The hair cuticle (Ce) consists of one to five or six layers (including overlapping) of cells. Each example in Fig. 3.9 consists of two cellular layers. They have granules, which as stated before are not thought to contain trichohyalin. As shown in Fig. 3.9A, most granules assemble in the outer (near-Ci) side of the cell, although some are still present at the center or inner side (see the cytoplasm of cell *b* in Fig. 3.9A). No tonofilaments are seen in these hair cuticle cells. The assembled granules enlarge and metamorphose to form a stone wall-like structure at the outer side of the cell, as shown in cell *a* of Fig. 3.9B. Some investigators call this wall the "exocuticle" [7] and the inner area the "endocuticle." The exocuticle develops as in cells *a* and *b* in Fig. 3.9C or 3.9D. The highly keratinized "A layer" [16], which should appear at the outside edge of the exocuticle, is seen (Fig. 3.10D), but this structure is frequently absent or indistinct in our samples. In Fig. 3.10D, the inside (distant from Ci) cell has a distinct A-layer whereas that in the outside Ce cell (adjacent to Ci) is ambiguous.

Extension of the exocuticle inward seems to be essential to the cornification of the Ce. The original spherical granules became scarce at this stage (see Fig. 3.9C). The outer cells (e.g., *a* in Fig. 3.9C) usually differentiate earlier than the inner cells (*b* in Fig. 3.9C), although this order is occasionally reversed. Eventually the cuticle cells (e.g., *a* in Fig. 3.9D) are closely packed with metamorphosed granules. Cornification of the cuticle starts later than the cortex, although note that on electron microscopy the density of the cuticle is remarkably higher than the cortex at the stage shown in Fig. 3.9D. Whether or not the inner-side granules are homogeneous with the wall-forming granules is an open question. In the initial phase of cornification of Ce cells, the distribution of the granules appears to be random (Fig. 3.10A), followed by subsequent accumulation in the outer side (Fig. 3.10B). I speculate that some granules migrate to the distal end of the cell by recourse to the cytoskeletal cables. The mechanism by which such polarity of Ce cells might be formed is not yet elucidated.

Fig. 3.9. Keratinizing process of hair cuticle. *Cx*, hair cortex; *Ce*, hair cuticle; *Ci*, cuticle of inner root sheath. Cells designated as *a* and *b* represent outer and inner layer of the cuticle, respectively. The cornification proceeds in alphabetical order (from **A** to **D**). A cell designated *a* in **D** is shown to have mostly accomplished the cornification. All samples were obtained from the dorsal skin of day 7 rats. *Bars* 500 nm

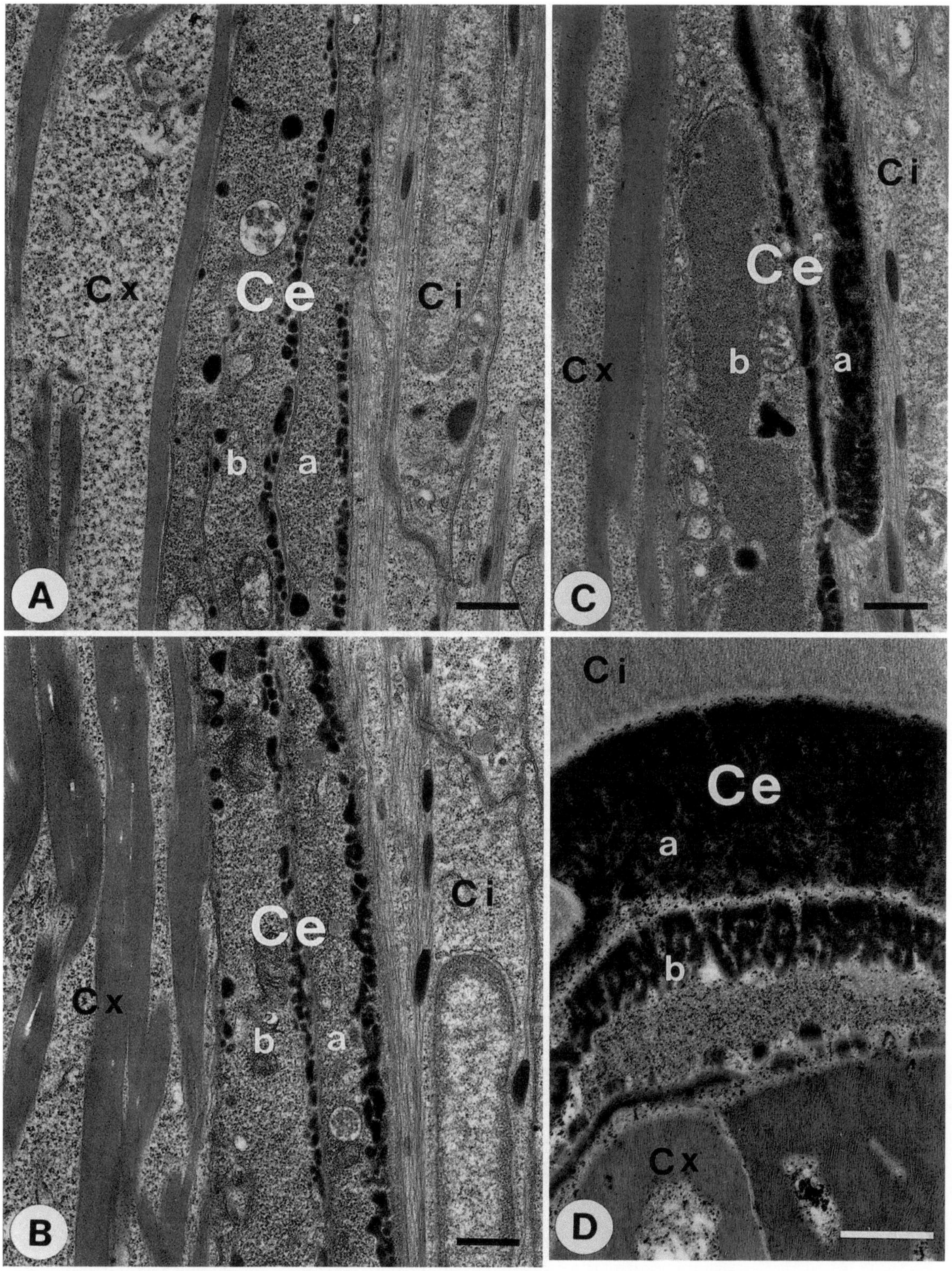

Fig. 3.9.

Granules

The finding of numerous amorphous or powdery granules in the narrow spaces between the exocuticular layers or in the precornification areas (one is denoted by a white arrow in Fig. 3.10D) is intriguing. These granules may be aggregates of denatured proteins under the process of digestion; alternatively, however, they may be the precursors of exocuticles. These amorphous granules are abundant in the Ce but are never seen in the Cx or Ci. If they were the products of lysosomal, proteasomal, or other proteolytic digestion, this appearance in the Ce only would be difficult to explain. Unlike the keratin bundles in the hair cortex, the metamorphosed and powdery granules never exhibit a fingerprint-like pattern. The multilayered tissue of the hair cuticle is shown in Fig. 3.10.

Details of the cuticle are shown in Fig. 3.10. Among these, Fig. 3.10A shows an almost random distribution of granules in the immature Ce, whereas Fig. 3.10C shows cornification occurring in the Ce layers from outside to inside. Figure 3.10E shows the special cornification of the hook area (see arrow) of the Ce that protrudes into the Ci. The inner side (Ce side) of the Ci is shown to start cornification in consonance with the formation of the exocuticular wall. This interlocking engagement between Ce and Ci ensures the synchronized growth of the hair shaft and inner root sheath until their departure (just before the emergence of the hair shaft) at the upper area near the skin surface.

Kvedar et al. have pointed out that the Ce shares certain characteristics with the stratum corneum of the epidermis, as the exocuticle in the Ce corresponds to the polar keratinization of the cornified envelope in the epidermis [16]. During the formation of the exocuticle, the cell membrane of Ce cells is reported to disappear [17]. In Fig. 3.9, Ce cells in A and B have cell membranes whereas those in Fig. 3.9C,D seem to have lost them. The structure of the boundary between the Ce and Ci and its neighborhood is discussed further in the section on the Ci.

Fig. 3.10. Multilayered hair cuticle and undetermined granules. *Cx*, hair cortex; *Ce*, hair cuticle; *Ci*, cuticle of inner root sheath. **A** An early cuticle cell before cornification. **B** Association of granules to the outer side of the cuticle cells. **C** Cornification of multilayered cuticle. **D** Enlarged view of the cornifying cuticle. The *white arrow* points to typical amorphous granules. Many similar granules are distributed in the interspaces of cornified layers. Some of them overlap upon the cornified areas. The granules tend to aggregate by themselves. **E** Special cornification at the boundary between hair shaft and inner root sheath (between Ce and Ci). The *arrow* denotes a "hook" area of a Ce cell. All samples were obtained from the dorsal skin of day 7 rats. *Bars* **A, B, C, E** 1 μm; **D** 100 nm

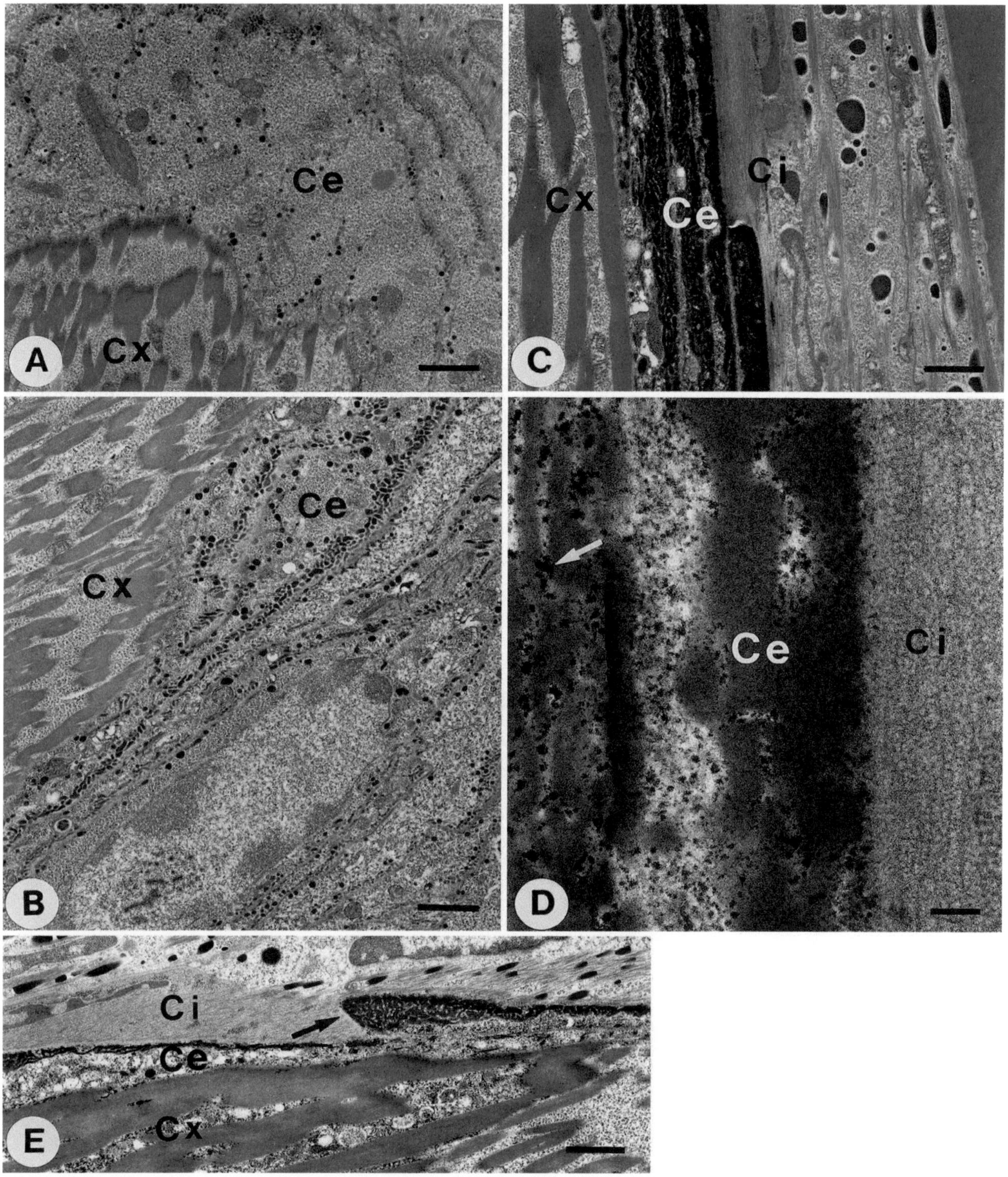

Fig. 3.10.

References

1. Marshall RC, Orwin DFG, Gillespie JM (1991) Structure and biochemistry of mammalian hard keratin. Electron Microsc Rev 4:47–83
2. O'Guin WM, Sun T-T, Manabe M (1992) Interaction of trichohyalin with intermediate filaments: three immunologically defined stages of trichohyalin maturation. J Invest Dermatol 98:24–32
3. Powell BC, Rogers GE (1997) The role of keratin proteins and their genes in the growth, structure and properties of hair. In: Jolles P, Zahn H, Hoker H (eds) Formation and structure of human hair. Birkhauser, Basel, pp 59–148
4. Morioka K, Takano-Ohmuro H, Sameshima M, Ueno T, Kominami E, Sakuraba H, Ihara H (1999) Extinction of organelles in differentiating epidermis. Acta Histochem Cytochem 32:465–476
5. van Leyen K, Duvoisin RM, Engelhardt H, Wiedmann M (1998) A function for lipoxygenase in programmed organelle degradation. Nature (Lond) 395:392–395
6. Takano-Ohmuro H, Mukaida M, Kominami E, Morioka K (2000) Autophagy in embryonic erythroid cells: its role in maturation. Eur J Cell Biol 79:759–764
7. Powell BC, Rogers GE (1997) The role of keratin proteins and their genes in the growth, structure and properties of hair. In: Jolles P, Zahn H, Hoker H (eds) Formation and structure of human hair. Birkhauser, Basel, pp 59–148
8. Parry DAD (1997) Protein chains in hair and epidermal keratin IF. In: Jolles P, Zahn H, Hoker H (eds) Formation and structure of human hair. Birkhauser, Basel, pp 177–207
9. Rhodin JAG (1974) Histology. A text and atlas. Oxford University Press, New York, p 174
10. Yang J, Liu X, Bhalla K, Kim CN, Ibrado AM, Cai J, Peng T-I, Jones DP, Wang X (1997) Prevention of apoptosis by Bcl-2: release of cytochrome c from mitochondria blocked. Science 275:1129–1132
11. Kroemer G (1999) Mitochondria control of apoptosis: an overview. In: Brown GC, Nicholls DG, Cooper CE (eds) Mitochondria and cell death. The Biochemistry Society, London, pp 1–15
12. Loeffler M, Kroemer G (2000) The mitochondrion in cell death control: certainties and incognita. Exp Cell Res 256:19–26
13. Morioka K, Tone H, Mukaida M, Takano-Ohmuro H (1999) The apoptotic and nonapoptotic nature of the terminal differentiation of erythroid cells. Exp Cell Res 240:206–217
14. Bassnett S, Mataic D (1997) Chromatin degradation in differentiating fiber cells of the eye lens. J Cell Biol 137:37–49
15. Jones LN, Horr TJ, Kaplin IJ (1994) Formation of surface membranes in developing mammalian hair fibres. Micron 25:589–595
16. Kvedar JC, Daryanani HA, Baden HP (1991) A comparison of the cross-linked components of stratum corneum of epidermis and the cuticle of the cortex of hair. The molecular and structural biology of hair. Ann NY Acad Sci 642:44–50
17. Lagermalm G (1954) Structural details of the surface layers of wool. Text Res J 24:17–25

Inner Root Sheath

4

Near the Skin Surface

The inner root sheath (IRS) consists of three concentric layers, the outermost "Henle's layer" (He), the intermediate "Huxley's layer" (Hx), and the innermost "cuticle of the inner root sheath" (Ci) (Fig. 1.4). Electron microscopic investigation of the structure of the IRS commenced with Birbeck and Mercer [1], who showed the transformation of trichohyalin granules in Henle's layer. The field of textile research considers the IRS to be less important than the hair shaft and it has not been well studied; even now its physiological meaning and details of its structure remain ambiguous.

The IRS escorts the hair shaft to the outside world, but does not form part of the emergent hair, because it is released from the shaft under the skin. Hair shafts do not seem to prick holes in the skin, but rather appear to be invited by the epidermis, where the IRS joins with the lower layers of the stratum corneum (cornified layers of epidermis), designated as number 1 in Fig. 4.1. Adhesion of the epidermal cornified layers to the IRS promotes the discharge of the IRS from the shaft, because the shaft extends while the IRS is entangled with the layers. The loci of adhesion are indicated by the arrowheads in Fig. 4.1. The structures around the hair shaft in Fig. 4.1B suggest that the separation of the IRS and ORS occurs before adhesion of the IRS to the cornified layers, although it remains possible that the separation (breakage) here resulted from the sectioning procedure itself. At the least, adhesion between the ORS and IRS seems to weaken considerably near the skin surface.

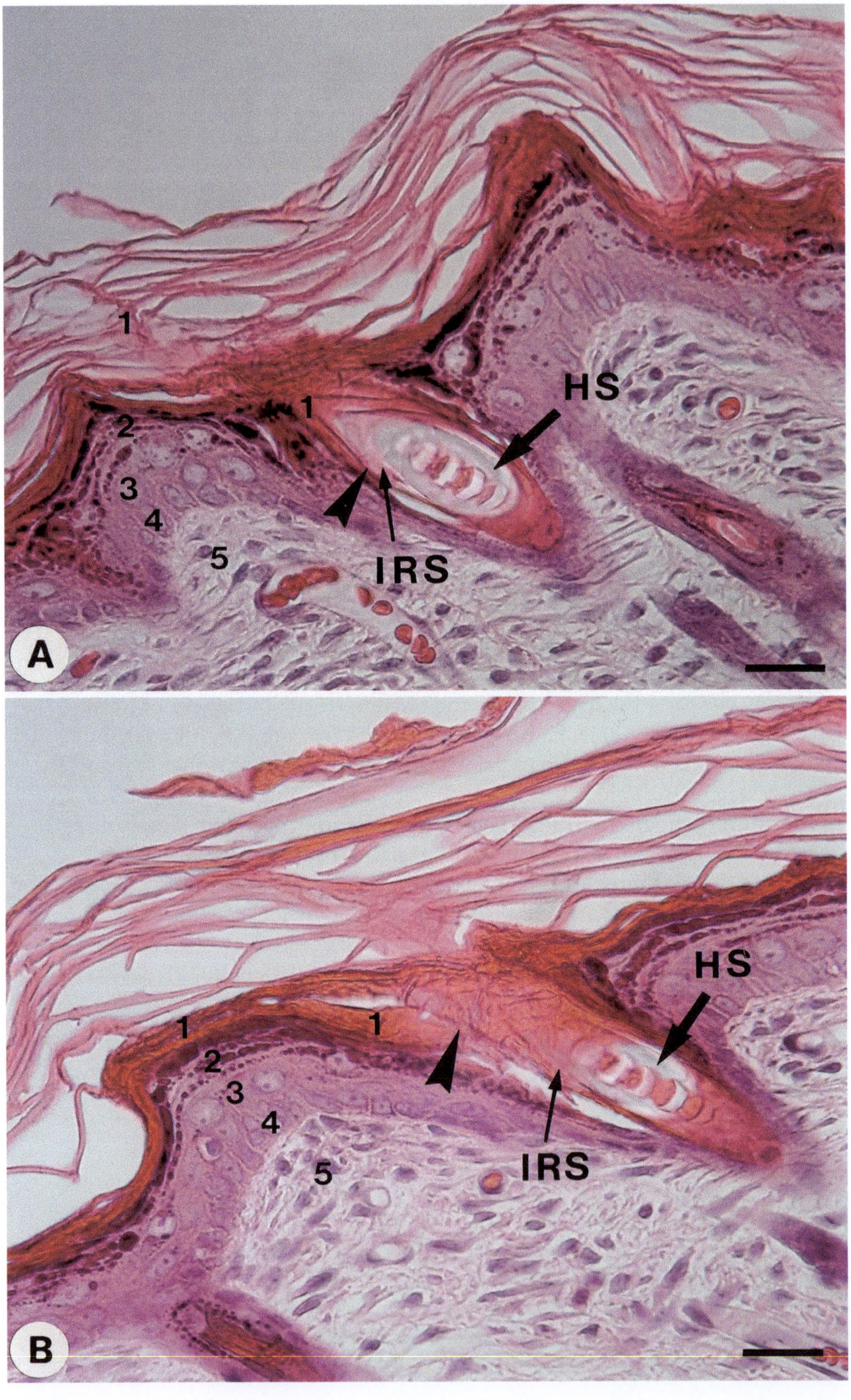

Fig. 4.1.

Precursors of the Inner Root Sheath

We now descend to the root area of the IRS to examine its developmental background. An oblique view of the hair bulb shows the cells near Auber's line (critical level) [2], below which the cells are not morphologically classifiable (Fig. 4.2). The differentiation advances from upper to lower in Fig. 4.2 (the top and bottom are reversed). In this figure, Henle's layer cells ahead of the black arrow contain a few tonofilaments and trichohyalin granules. The asterisk-marked dividing cell at the upper end of the figure is expected to be a monopotent stem cell destined to produce Henle cells, or, in other words, it is the root of Henle's layer. Cells at the left of the asterisk-marked cell (ahead of the open arrow) show no sign of differentiation, suggesting that they may be immature stem cells. The "stem cell wheel" (see Fig. 6.11, later) may be composed of these cells.

Compared with the pelage hair, the vibrissae of animals grow very rapidly, as fast as 1 mm per day in rats, for example. Even human scalp hairs grow 0.3 mm or so. A length of 1 mm corresponds to about 100 cells or more. To maintain this rate, a number of basal cells are needed for one series of cells. They are arranged in tandem because a single basal cell is unable to produce 100 cells per day. On the basis of this arrangement, it is expected that the dividing basal cell rolls aside after division and a new cell migrates into its position, with this process repeating over and over. Hypotheses explaining the alternation of basal cells are shown in Fig. 6.11. The deduced boundary between the hair matrix and root sheath precursors is shown by the dotted line in Fig. 4.2. The area above the upper end of the dotted line in Fig. 4.2 is filled with undifferentiated hair matrix cells, some of which are considered to be multipotent stem cells. Although it has been demonstrated that the stem cells are stored in the bulge area in adult animals [3–5], neonatal animals need to retain multipotent stem cells in their hair matrix because the bulge has not been formed at this stage. It is known that adult follicles have secondary hair germs in the hair matrix, in addition to the primary niche of stem cells in the bulge area [6].

Fig. 4.1. Penetration of hair shaft through the epidermis. **A, B** Histochemical views around the penetration point of the hair shaft: *1*, cornified layer; *2*, granular layer; *3*, spinous layer; *4*, basal layer; *5*, dermis. Each *arrowhead* indicates the position of the adhesion of cornified layer of the epidermis with the inner root sheath (*IRS*), which is designated by the *thin arrow*. *Thick arrow* denotes hair shaft (*HS*). Samples were obtained from the dorsal skin of day 2 rats. The paraffin sections were stained with the conventional hematoxylin and eosin (H&E) procedure. *Bars* 20 μm

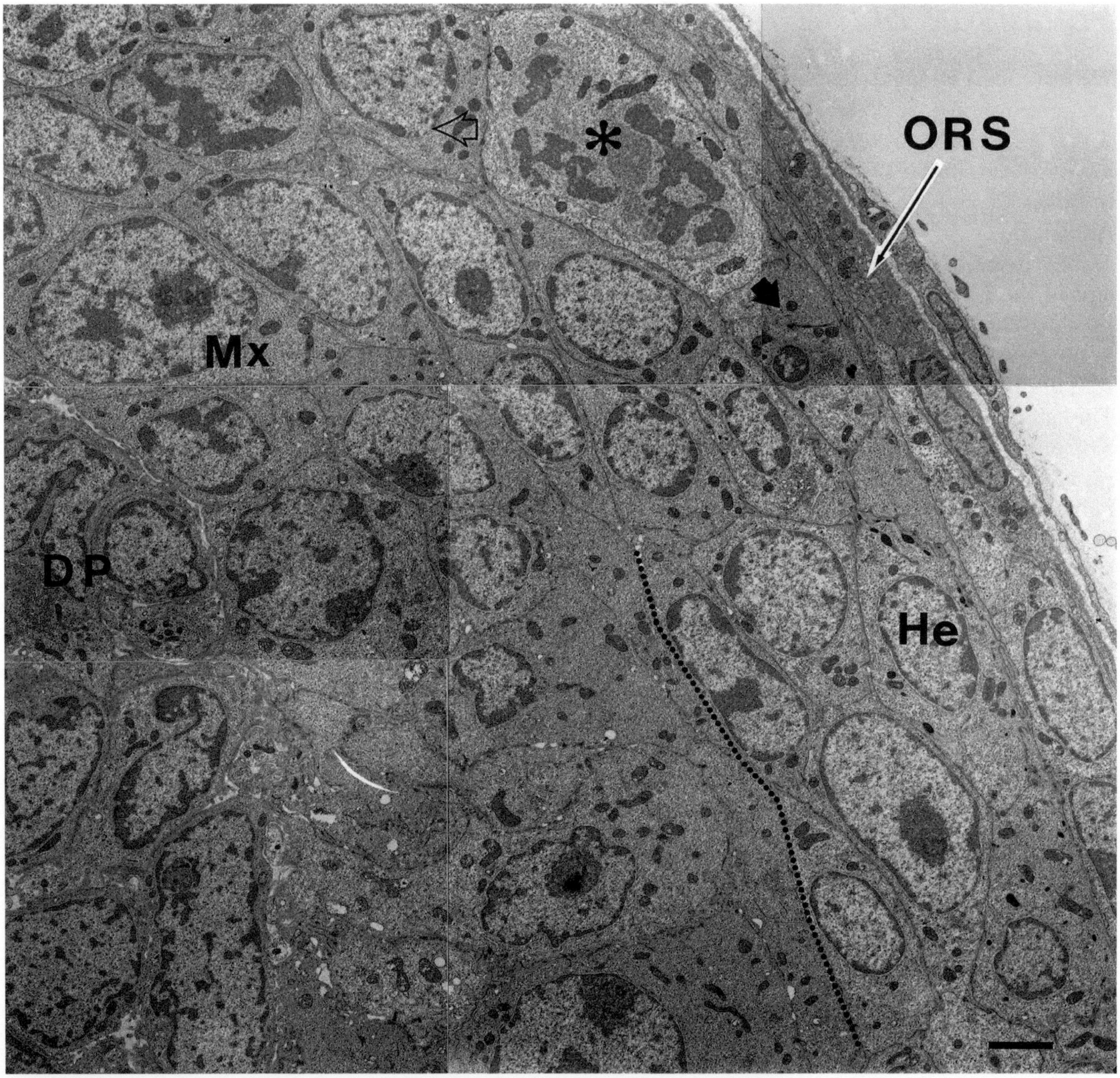

Fig. 4.2. Intermediate height of the hair bulb. *ORS*, outer root sheath; *He*, Henle's layer; *Mx*, hair matrix; *DP*, dermal papilla; *, predicted root cell of Henle's layer. *Black arrow*, direction of the hair growth and differentiation; *open arrow*, area of immature cells; *dotted line*, predicted boundary between precursors of inner root sheath cells and hair matrix cells. Samples were obtained from facial skin of day 2 rats. *Bars* 1 μm

Formation of Trichohyalin Granules and Tonofilaments

Figure 4.3A shows the early IRS, in which some trichohyalin granules with sparse tonofilaments are seen in the He layer. By contrast, the Hx and Ci do not exhibit significant signs of differentiation. The electron density of the cytoplasm of the IRS layers is less than that of the hair shaft layers at this height. When the stage advances to that in Fig. 4.3B, the three layers of the IRS enter into the process of keratinization and are easily identifiable by morphology. Trichohyalin granules are not yet evident in the Ci cells in Fig. 4.3B, but do appear in the later stages, as shown in Fig. 3.9A,B, for example. He and Ci subsequently progress to full keratinization while this process is still underway in Hx (Fig. 4.3C). The delay of keratinization in Hx may be responsible for retaining the plasticity of the sheath until just before the release of the hair shaft, and at the same time may benefit the hair follicle with regard to informational transfer or nutrition, as Hx is the only "living" layer in the last stage of hair differentiation. Eventually, Hx is also fully cornified (Fig. 4.3D). At around this time, the hair shaft is thought to be released from the internal root sheath (Fig. 4.3D), although a precise determination of this point is not possible because the release of the shaft may be due to an artifactual effect during the production of ultrathin sections (Figs. 4.3D, 4.5).

Special Adherence of Henle's Layer Cells

During their differentiation, He cells form a specific jagged junction structure of cell membranes at the intercellular junctions between them (very thick arrow in Fig. 4.4C). This boundary includes many desmosomes, one of which is highlighted by the thin arrow in Fig. 4.4C. By contrast, the boundaries between He–Hx or He–Cp (Cp, companion layer) cells are largely smooth, with scarce desmosomes (Fig. 4.4A–C). Because He layer cells keratinize earlier than other inner layer cells, they may lose the potential to divide or grow. They must cope with expansion pressure generated from the inner follicle, which is brought about by the continual growth and division of hair matrix cells. The strong lateral adhesion between He cells may therefore be a means by which they endure this pressure. He cells under this pressure create a special tight binding system (Fig. 4.4C) and become thin. The jagged junction structure shown in Fig. 4.4C is not a rare case, but rather is commonly observed at He–He boundaries (see also Fig. 4.6) [7].

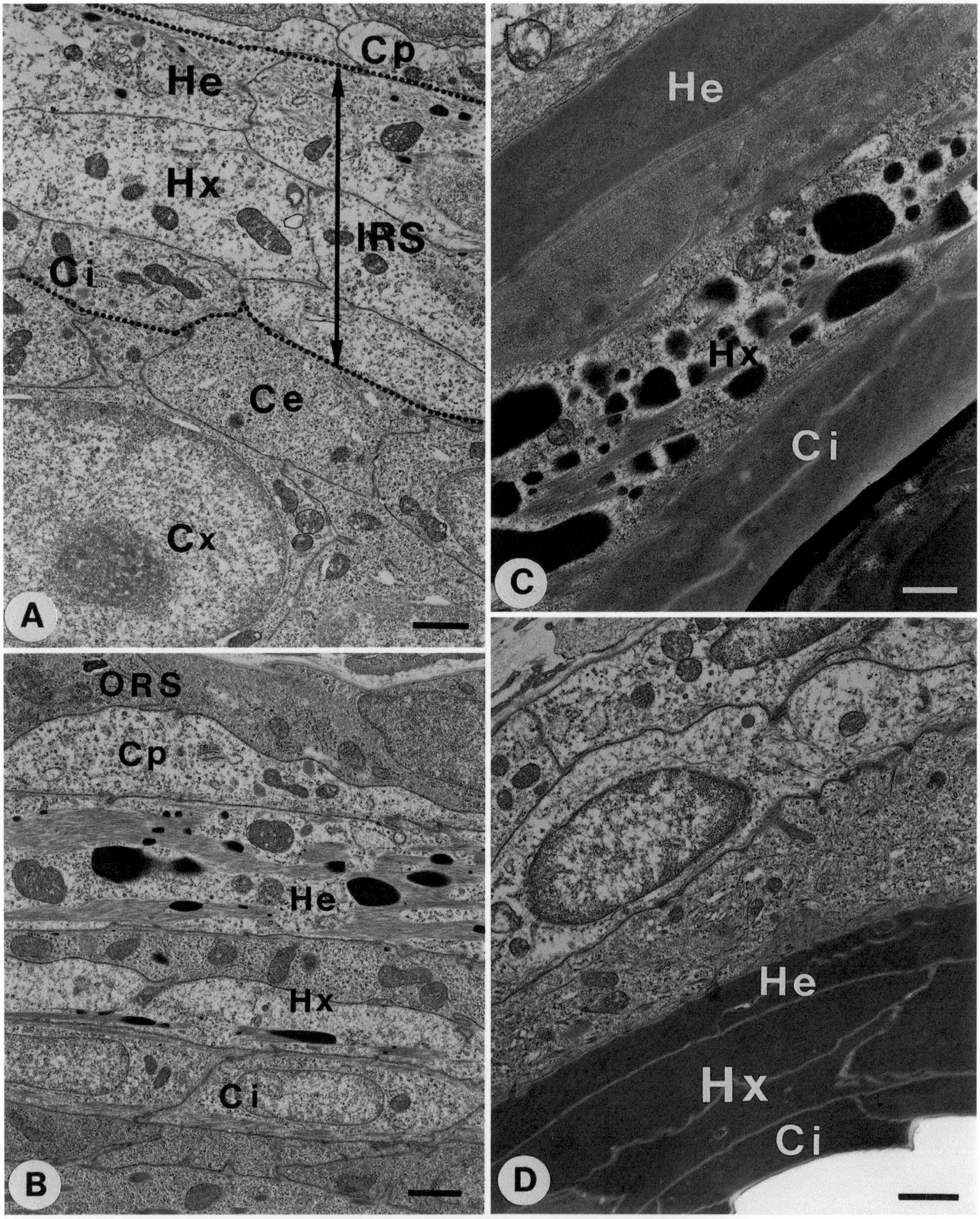

Fig. 4.3. Differentiation of inner root sheath cells (*outline*). Maturation of the cells proceeds from **A** to **D**. *ORS*, outer root sheath; *Cp*, companion layer; *He*, Henle's layer; *Hx*, Huxley's layer; *Ci*, cuticle of inner root sheath; *IRS*, inner root sheath; *Ce*, hair cuticle; *Cx*, hair cortex. The area lying between the *two dotted lines* in **A**, designated by the *vertical bar with dual arrowheads*, corresponds to immature IRS. Samples were obtained from dorsal skin of day 7 rats. *Bars* **A**, **B**, **D** 1 μm; **C** 500 nm

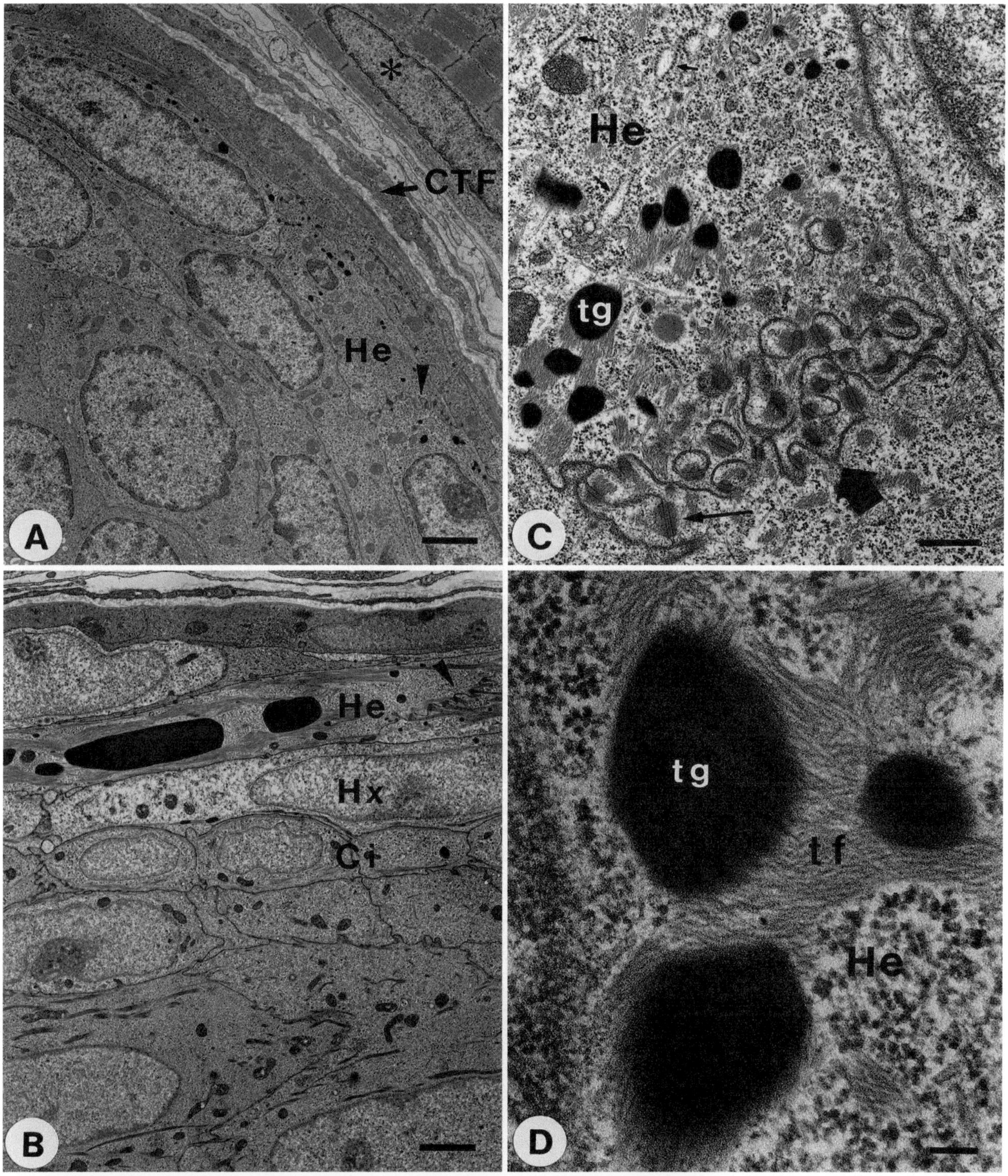

Fig. 4.4. Differentiation of Henle's layer: granules, filaments, and special junctions. *CTF*, connective tissue follicle; *He*, Henle's layer; *Hx*, Huxley's layer; *Ci*, cuticle of inner root sheath; *tg*, trichohyalin granule; *tf*, tonofilament. *Asterisk* in **A** denotes a dermal muscle cell. *Small arrows* in **C** (*upper, left*) indicate rough endoplasmic reticulum; *thick arrow* denotes the boundary of the cells of Henle's layer; *thin long arrow* in **C** (*lower, center*) indicates one of the desmosomes. There are numerous desmosomes on the boundary between Henle's layer cell and its adjacent Henle's layer cells. Samples were obtained from facial skin of day 2 rats. *Bars* **A**, **B** 2 μm; **C** 0.5 μm; **D** 0.1 μm

Decomposition of Trichohyalin Granules and Cornification

The tiny arrows in the upper areas of Fig. 4.4C denote rough endoplasmic reticulum, which seems to be swollen but are not yet destroyed. If the swelling is in fact a sign of future degeneration, intracellular metabolism must be very active at this stage of differentiation. Tonofilaments (tf) are often associated with trichohyalin granules (tg) (Fig. 4.4D). Although it is not presently thought that tonofilaments are produced from loosened trichohyalin granules [8, 9], Figs. 4.4D and 4.5B indicate a close relationship between filaments and granules. Trichohyalin molecules are likely involved in the assembly of keratin-based fibers.

Figure 4.5 shows the morphology of He cells with special regard to the degradation of trichohyalin granules in the critical period of keratinization. Amorphous structures of trichohyalin granules are depicted by the thin arrows in Fig. 4.5B (early) and 4.5A (late). They seem to be in the process of degradation. Fully cornified cells have no granules in their cytoplasm (Fig. 4.5C). Because the degradation of trichohyalin granules proceeds synchronously and rapidly, it is no easy matter to capture them in actual degradation. I was fortunately able to capture one image that shows a significant overlap between granules and fibers (Fig. 4.5B). That is, both are present or both are absent in each limited area of attention. Similar images have appeared previously (e.g., fig. 17 in Reference 10) . As shown in cells 1 through 4 in Fig. 4.6, it is usually possible to observe only a decrease in granules, as in cells 2 and 3, or the abrupt appearance of a cornified state, as in cell 4. The rapid progression of special and critical differentiation processes of living cells is an ongoing problem for morphologists, especially in the field of dermatology. The photographs in Figs. 4.5 and 4.7 showing an intermediate state are therefore particularly valuable.

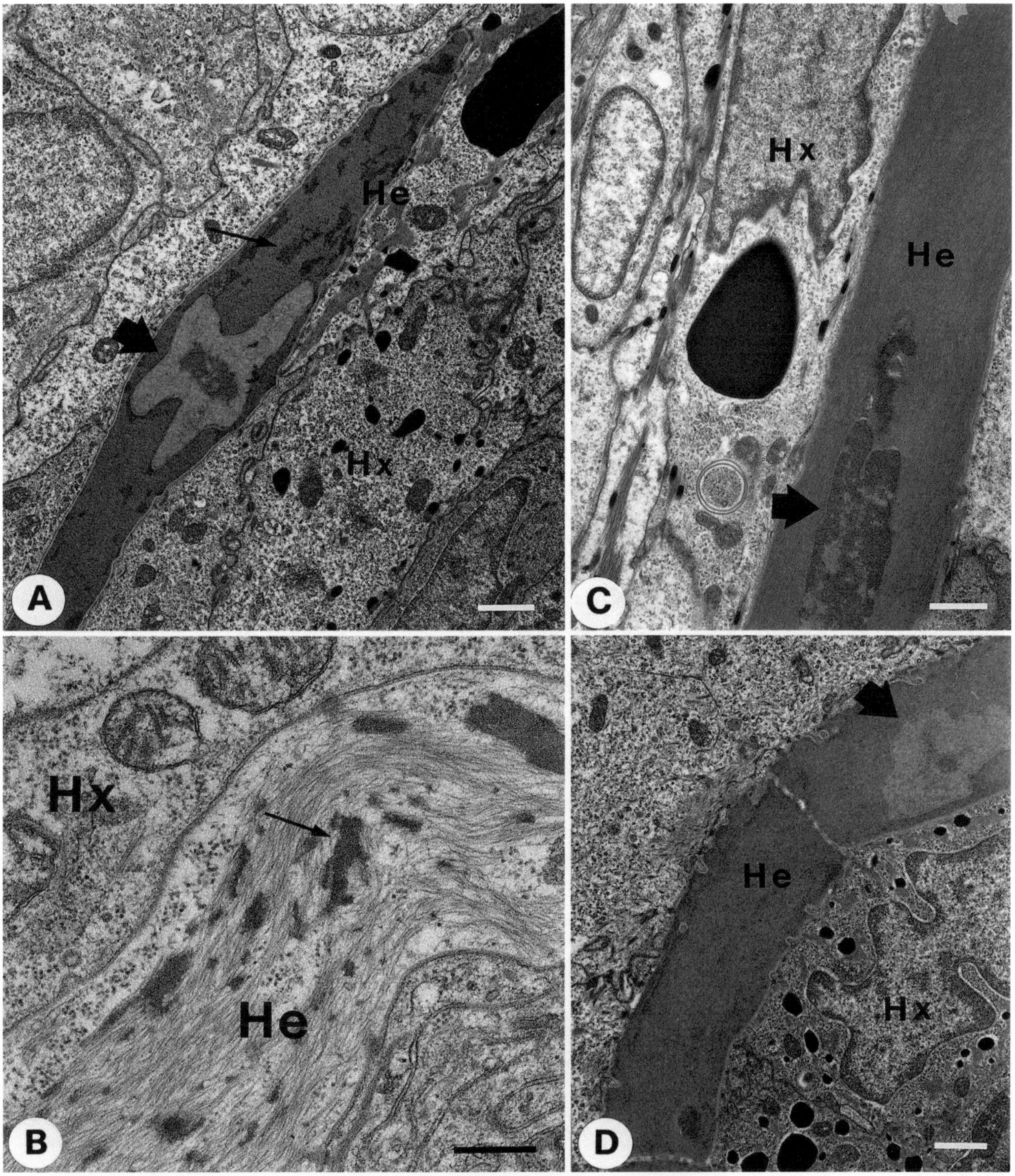

Fig. 4.5. Cornification of Henle's layer. Degradation of granules and nuclei. *He*, Henle's layer; *Hx*, Huxley's layer; *thin arrow*, deformed granule; *thick arrow*, degenerated nucleus. Samples were obtained from facial skin of day 2 rat (**A**), dorsal skin of day 7 rat (**B**), dorsal skin of day 7 rat (**C**), and facial skin of day 2 rat (**D**). Bars **A**, **C**, **D** 1 µm; **B** 500 nm

The He cells just before cornification are shown in Fig. 4.7. As stated in the section on the medulla, trichohyalin molecules, which are not included as components of the filaments, cornify the cell by forming isopeptide bonds with one another. In contrast to the medulla, IRS cells produce a great deal of keratin-based intermediate filaments in the process of cornification, as shown in Fig. 4.7. This process, in other words, is keratinization. Figure 4.7A,B shows that keratin filament bundles form several clusters in the cell. According to previously published results [11–13], filament production is suggested to be efficient at sites in the vicinity of trichohyalin granules. This locally enhanced production of filaments seems to involve the clustering of keratin filaments shown in Fig. 4.7. Nevertheless, it is evident that the granules themselves are not needed for the increased production of filaments required to cornify cells, as they seem to disappear before the critical period (Fig. 4.7A,C). On the other hand, the question remains as to why at least some degenerating granules may still be seen in some cornified cells (Fig. 4.5A; see also Fig. 4.10D). Clearly, the mechanism of keratinization is still obscure.

In most cases, follicle cells provide no hint of the final critical stage of cornification. This finding indicates the extremely high speed accumulation, assembly, or both of keratin filaments in this stage, which may be induced by components of trichohyalin granules. In contrast to the keratinization of the IRS, that of the HS, which usually has no trichohyalin granules, is distinctly slow, and consequently the process of keratin filament accumulation is detectable (notwithstanding the technical difficulties arising from its refractoriness to the penetration of embedding resins).

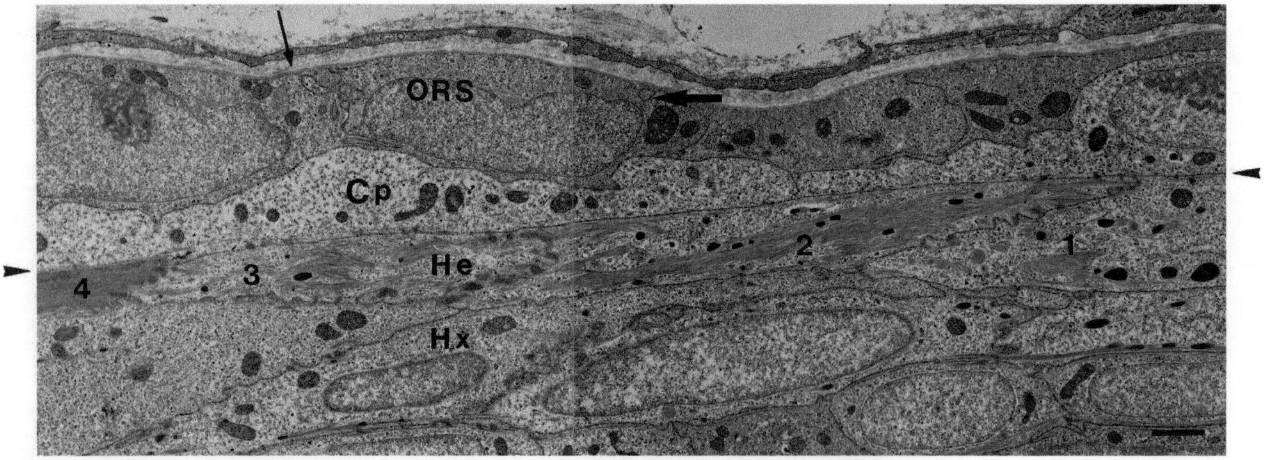

Fig. 4.6. Inner and outer root sheath cells. Boundary of the cells. Henle's layer cells undergoing differentiation. *ORS*, outer root sheath; *Cp*, companion layer; *He*, Henle's layer; *Hx*, Huxley's layer. *Horizontal thick arrow* represents the direction of skin surface; *thin arrow* indicates the space between dermal and epidermal (ORS) cells. *Arrowheads* at both sides designate the border between Cp and He. Doral skin from a day 7 rat was used. *Bar* 1 μm

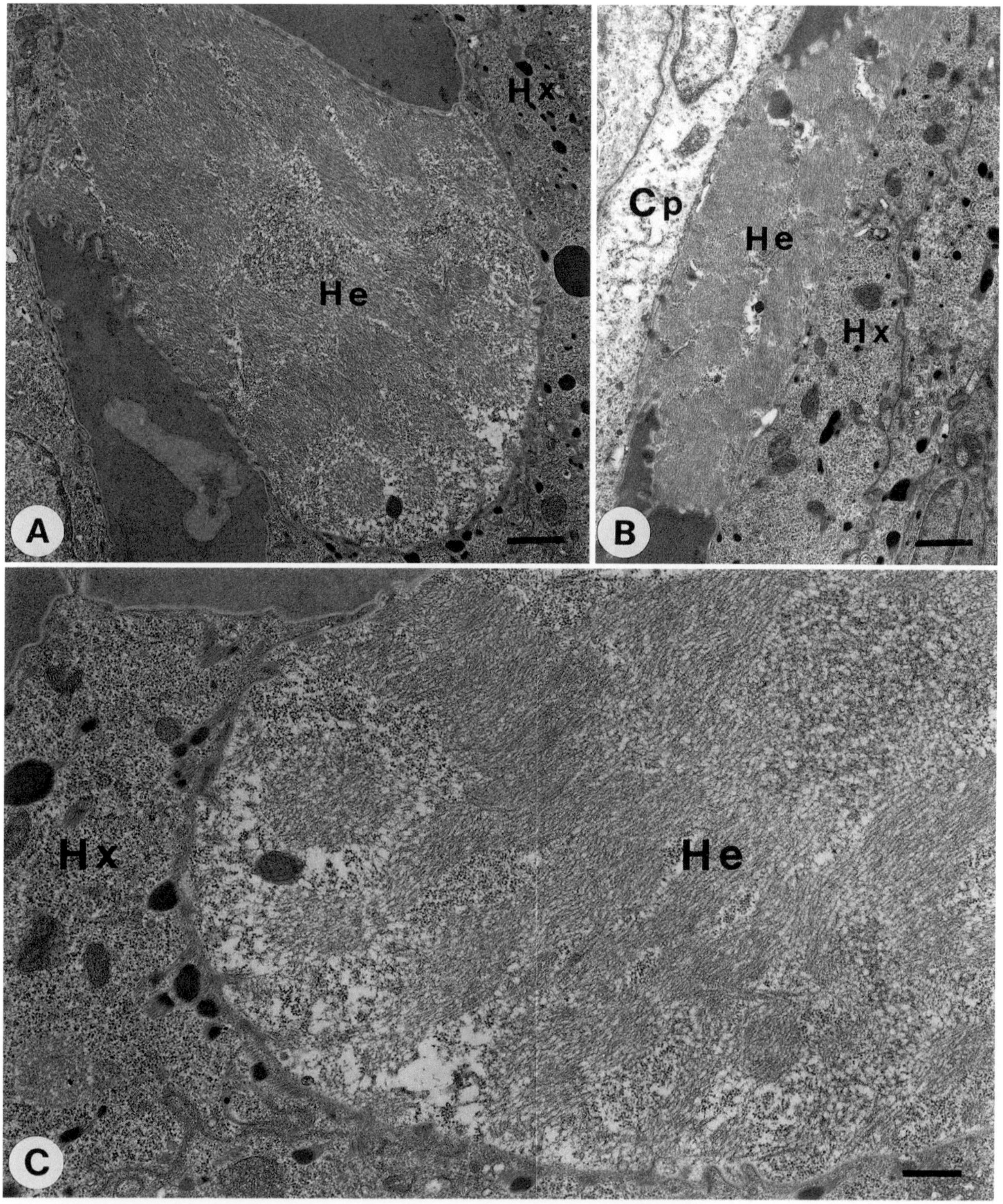

Fig. 4.7. Henle's layer cells just before the cornification. *Cp*, companion layer; *He*, Henle's layer; *Hx*, Huxley's layer. Samples were obtained from facial (**A**, **C**) and dorsal (**B**) skin of day 2 rats. **C** Enlarged view of a part of **A**. *Bars* **A**, **B** 1 μm; **C** 500 nm

Cellular Junctions

Figure 4.6 shows not only the differentiation of He but also several other points of note. The nearly straight horizontal line marking the border between the outer and inner root sheaths (see arrowheads) resembles tissues of the hair shaft in the early stage (see Fig. 3.2). This simple line contrasts with the other curved and rugged boundaries of the cells, suggesting that epidermal cells possess several other types of cellular junction in addition to those previously classified on the basis of the length of the space between the membranes or the presence of special apparatus such as desmosomes or connexons. I expect to see further developments in the study of tight junctions and the function of claudin family proteins in hair [14].

The thin arrow in Fig. 4.6 denotes a space equivalent to the basement membrane between the skin epidermis and dermis. Epidermal outer root sheath cells and dermal cells are separated by a 100- to 500-nm gap, which is probably filled with extracellular matrix proteins such as collagens and laminins. Mitochondria in He are shown to be lost before its cornification. The mechanism of this loss (or degeneration) remains to be determined.

Huxley Versus Henle Versus Cuticle

The differentiation of Hx differs from that of He in some respects. (1) He cornifies first whereas Hx is the last to do so among all hair follicle tissues. (2) In Hx cells, the spherical trichohyalin granules become amorphous and accumulate to fill the whole cytoplasm just before cornification (Fig. 4.8D). Accumulation of granules is usually not seen in He. (3) Tonofilaments accumulate in He as differentiation proceeds, and can also be seen in Hx from the early stages of differentiation (Figs. 4.3B, 4.8A); the presence of packed trichohyalin granules in Hx cells suggests that the cross-linking of trichohyalin may play a significant role in the cornification of Hx layers (Fig. 4.8B,D). The stone wall-like (exocuticle-like) structure in Fig. 4.8D apparently resembles the cornification of the hair cuticle, but unlike the cuticle, these structures do not fuse with each other to form a cornified cytoplasm. The cornified cytoplasm of Hx cells is fibrous, as shown in Figs. 4.8D, 4.9D, and 4.15 (a magnifying glass may be useful). The amorphous special granules that fill the gaps between the granules of the hair cuticle are not found in Hx, but remnants of trichohyalin granules are seen among newly accumulated tonofilaments in the cornifying Hx (Fig. 4.8C,D). Although Hx cornifies at last, the formation of the graules in Hx starts earlier than the appearance of differentiation signs in Ci.

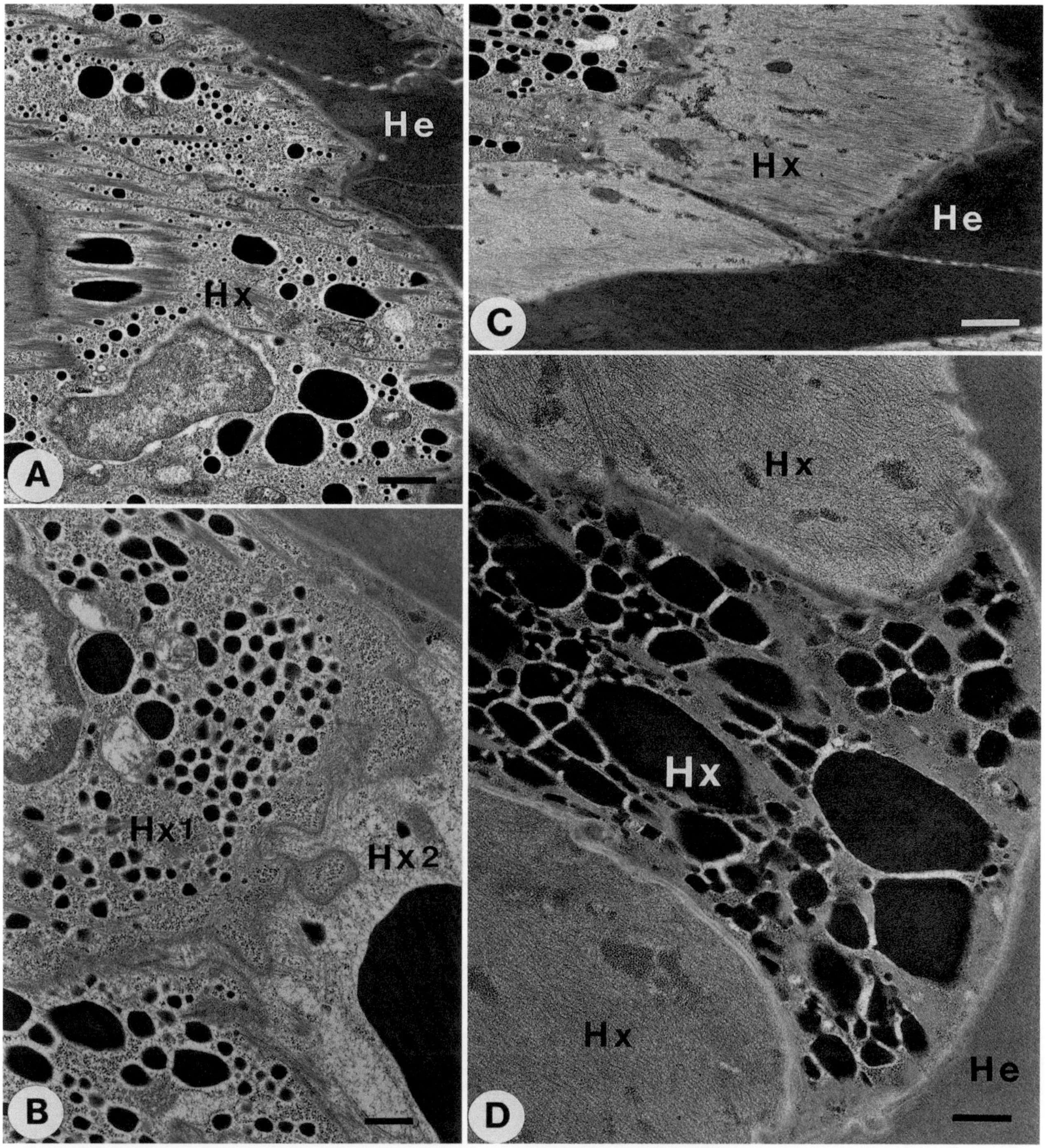

Fig. 4.8. Huxley's layer cells undergoing cornification. *He*, Henle's layer; *Hx*, Huxley's layer. *Hx 1* seems to be nearer to the final cornification than *Hx 2*. Dorsal skins of day 7 (**A**, **D**) and day 2 (**B**, **C**) rats were used for the experiment. *Bars* **A**, **C** 1 μm; **B**, **D** 500 nm

Degeneration of the Nucleus

The morphological changes of nuclei may in large part be common among tissues of the inner root sheath. The nuclei of He cells become transiently electron lucent in comparison to the surrounding cytoplasm, as denoted by the thick arrows in Fig. 4.5A,D. At the same time, electron-dense pyknotic nuclei including condensed chromatin can be seen (thick arrow in Fig. 4.5C). Whether these morphological changes are consecutive or independent has not been determined.

The process of nuclear degeneration in Hx cells seems closely similar to that in He. Figure 4.9A shows Hx cells in the early stage of differentiation. Several small granules and some tonofilaments already appear in these Hx cells, whose nuclei are largely spherical (Fig. 4.9A). As differentiation proceeds the nuclei become amorphous or pyknotic (Fig. 4.9B,C), but I cannot yet determine whether the next step of degeneration is as in Fig. 4.9D or as in Fig. 4.9E. The former nucleus seems to contain dispersed chromatins, while the latter is filled with highly electron-dense chromatins. Even at the stage seemingly just before the complete degradation of the nucleus, both types of nuclei are found, as designated with a black "n" and a white "n" in Fig. 4.9F. The deformation is more extreme in the electron-dense type of nucleus. For example, rodlike protrusions in an Hx cell are shown in Fig. 4.9E. This type of nucleus in He cells also tends to undergo extreme deformation.

Degeneration of the nucleus during keratinization of the Ci is also shown. The condensed chromatin is initially associated with the nuclear periphery, as in the case of apoptosis (see Fig. 4.14A). The nuclear membrane then becomes obscure (arrows in Fig. 4.14B–D), however, after which the chromatin becomes electron lucent (thick arrow in Fig. 4.14D). This pattern of degradation differs from typical apoptotic nuclear degeneration.

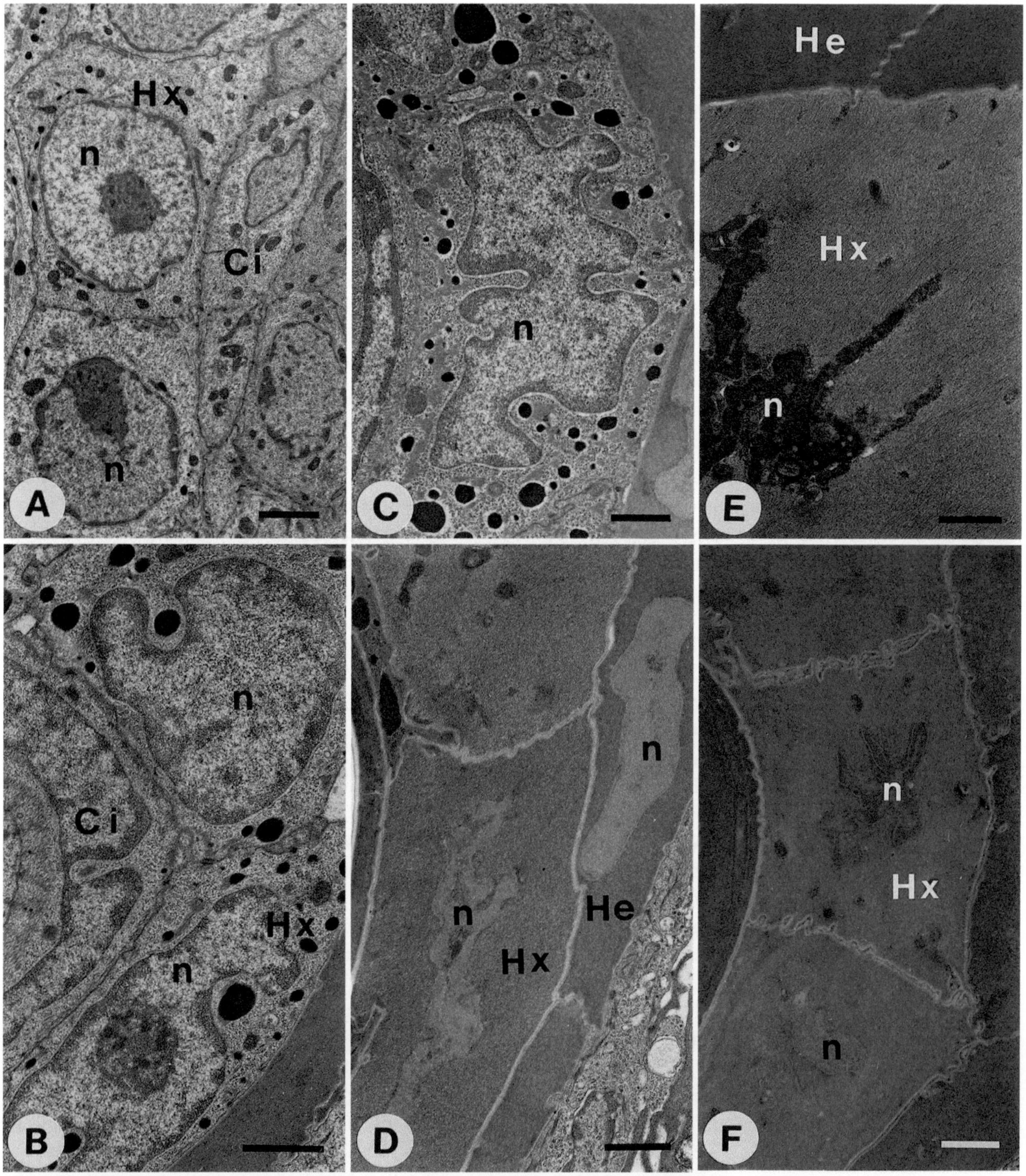

Fig. 4.9. Degeneration of nucleus of Huxley's layer cells. The differentiation proceeds from **A** to **F**. *He*, Henle's layer; *Hx*, Huxley's layer; *Ci*, cuticle of inner root sheath; *n*, nucleus. Facial and dorsal pieces of skin obtained from day 2 rats were used in **A–C** and **F**, respectively; dorsal skin of a day 7 rat was used in **D** and **E**. *Bars* **A** 2 μm; **B–F** 1 μm

Specifics of the Cuticle of the Inner Root Sheath

The process of cornification in the cuticle of the inner root sheath (Ci) is different from that in the other two layers of the IRS with regard to the polarity of keratinization. The Ci layer adjoins the Hx (Huxley's layer) at its outer surface and the Ce (hair cuticle) at its inner surface. Keratinization of the Ci starts mostly from the Ce (inner) side, as indicated by the arrowheads in Fig. 4.10A,C, although keratin fibers do not attach to the cell membrane of the Ce side. A space is usually present (asterisks inserted between keratin fibers and cell membrane in Fig. 4.10B,D) between the fiber bundles (thick arrows in Fig. 4.10B,D) and cell membrane. As differentiation proceeds, keratin fibers reach the Ce side of the cell membrane of the Ci. At that time, trichohyalin granules emerge in the Ci (Fig. 4.11), as in the case of the other two (He and Hx) layers (see Fig. 4.10A–D). The order of appearance of the granules is first He, followed by Hx, and finally Ci. This observation is consistent with the previous result for the order of appearance of tonofilaments, which is He, Hx, and then Ci [15], but not with that of cornification, which is He, Ci, and then Hx.

The contact of fibers in the Ci with its Ce-side membrane appears to be correlated or synchronized with the keratinization of the Ci side of Ce (Fig. 4.11C,D). Furthermore, the inner ends of the fibers seem to combine with the exocuticle of the Ce across the Ci–Ce boundary. It may be important for maintenance of the rigidity of the Ce side of the Ci that it collaborates with the exocuticle of the Ce to result in a strong but removable junction. The significance of "strong but removable" is discussed in Fig. 4.16. Nevertheless, the growth of the fibers does not directly depend on Ce cornification, because I found a case in which no accumulation of granules or exocuticle was found in the Ci side of the Ce (near the dotted line in Fig. 4.11A). The association of Ci fibers with the exocuticle is probably effective in determining the localization and direction of the fibers. The tonofilaments in Fig. 4.12D (denoted by the arrow) seem to penetrate the cell membrane of the Ci and extend to contact with the exocuticle of the Ce. Figure 4.12C,D shows that there is a margin between the exocuticle and cell membrane of the Ce, although the physiological significance of this is unknown.

The mechanism by which the polarity of the Ci is formed is not yet known. One clue is that trichohyalin granules in the Ce side are of smaller average size than those of the Hx side (Fig 4.12A–C). When comparing the size distribution of trichohyalin granules, I was surprised to find that the outer and inner sides of the Ci cell in Fig. 4.12A–C could be partitioned by a white dotted line drawn in those figures. One explanation is that the small granules are in the process of degradation, and that the degradation products induce the formation of tonofilaments. The granules in the area rich in fibers are heterogenous (some are just fading) and smaller (Fig. 4.12C,D), an appearance which resembles that in the keratinization process of Hx or He.

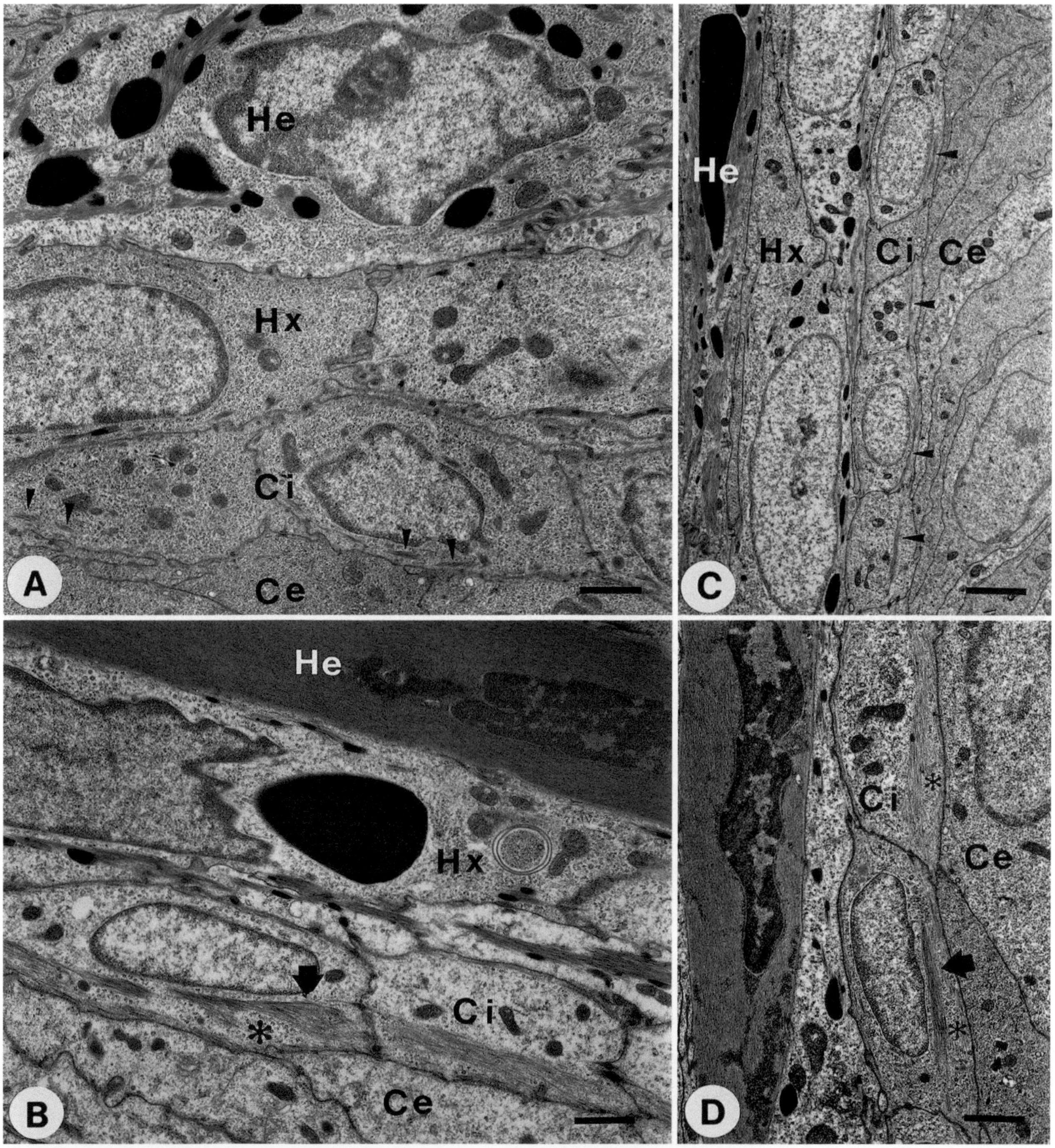

Fig. 4.10. Formation of tonofilaments in cuticle of inner root sheath during the early stages. *He*, Henle's layer; *Hx*, Huxley's layer; *Ci*, cuticle of inner root sheath; *Ce*, hair cuticle. *Arrowheads* in **A**, **C**, and *arrows* in **B**, **D** denote the tonofilaments formed at the Ce side of Ci; *asterisks* in **B** and **D** represent the space between tonofilament bundles and cell membrane in Ci cells. **A** Facial skin of day 2 rat; **B–D** dorsal skin of day 7 rats. *Bars* **A**, **B**, **D** 1 µm; **C** 2 µm

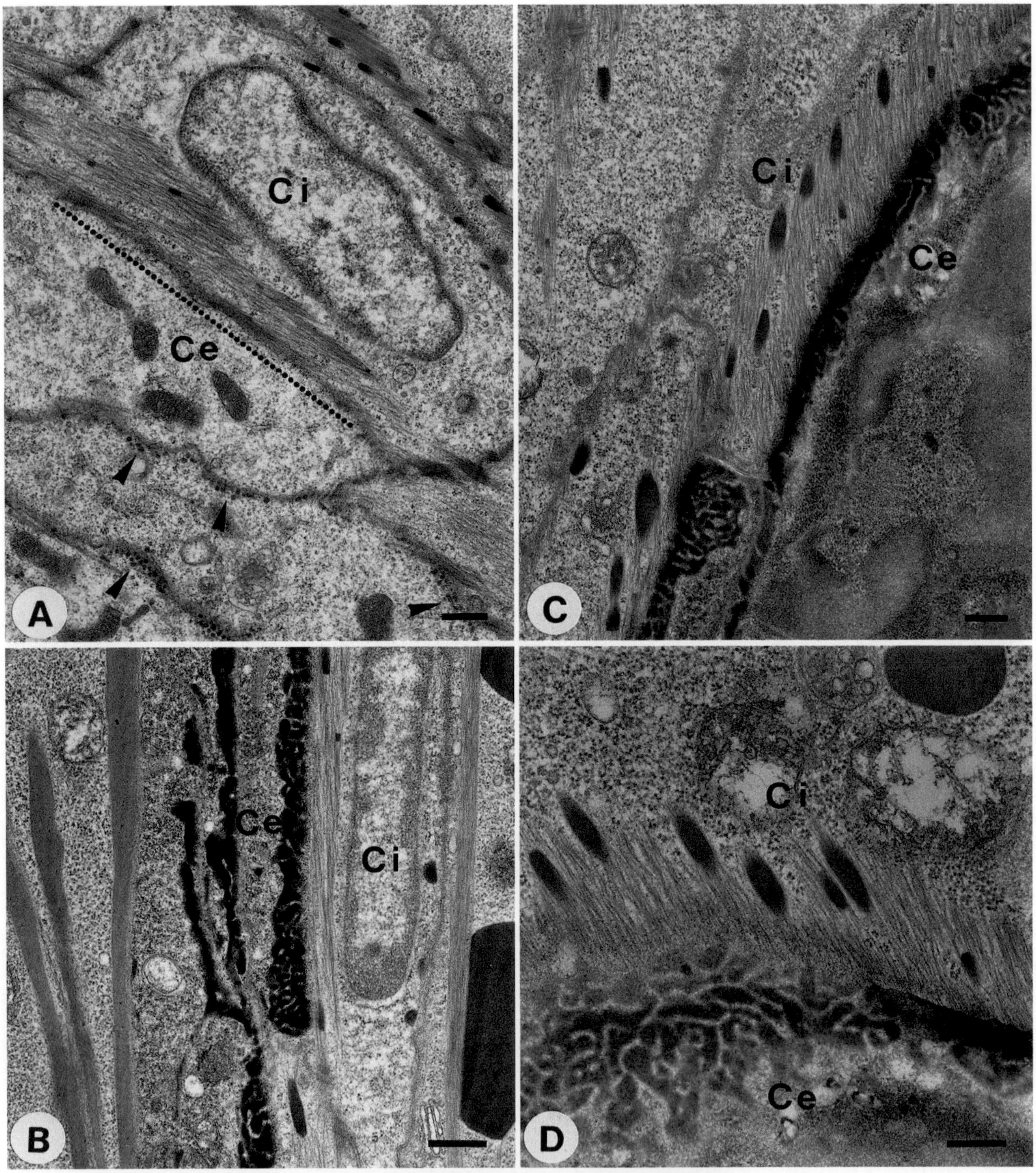

Fig. 4.11. Localized cornification in the cells of cuticle of inner root sheath. *Ci*, cuticle of inner root sheath; *Ce*, hair cuticle. *Arrowheads* in **A**, immature granules in the *Ce* cells; *dotted line* in **A**, a border area of Ce that is facing Ci. The immature granules are scarce in the area near the *dotted line*. Dorsal skin obtained from a day 7 rat was used for every experiment. *Bars* **A–C** 500 nm; **D** 200 nm

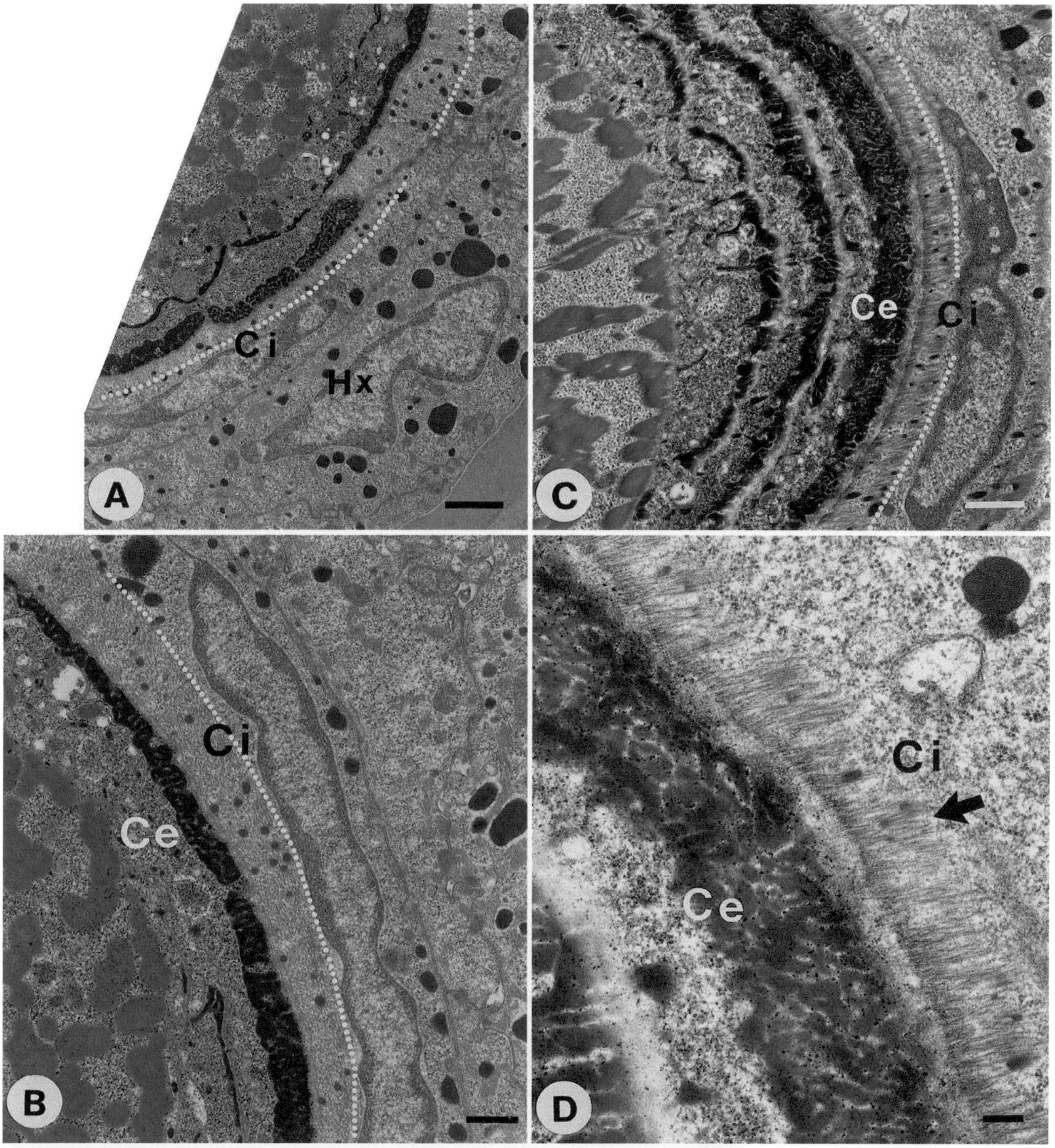

Fig. 4.12. Polarity of cornification in the cytoplasm of inner root sheath cells. *Hx*, Huxley's layer; *Ci*, cuticle of inner root sheath; *Ce*, hair cuticle. The Ci cells have two distinct areas in their cytoplasm during the course of the differentiation; one is rich in tonofilaments while the other is scarce in filaments. The border is designated by *white dotted lines* in **A**, **B**, and **C**. The *arrow* in **D** denotes the appearance of the tonofilaments at the near-boundary area of Ci that is adjacent to Ce. **A** Facial skin of day 2 rat. **B** Dorsal skin of day 2 rat. **C** Dorsal skin of day 7 rat. **D** Dorsal skin of day 7 rat. *Bars* **A** 1 μm; **B** 500 nm; **C** 1 μm; **D** 500 nm

Relationship Between the Cuticle of the Inner Root Sheath and the Hair Cuticle

In Fig. 4.13, the hair shaft is still at an early stage of development and exhibits no sign of cornification, whereas He has already cornified, as shown in part A. The Ci and Ce are often connected by numerous desmosomes at this stage (Fig. 4.13B). In later stages, the junction of Ci and Ce forms a special wedgelike structure (Figs. 3.4B, 3.7E, 3.10C, 4.11A–C), usually without the aid of desmosomes, but occasionally these are retained even after cornification of the Ce and Ci (see the border areas of Ce and Ci in Fig. 4.14C). Filament formation in the Ci is observed around the desmosomes. It is possible to partition the Ci cell into filament-rich and -poor areas, as shown by the white dotted line in Fig. 4.13B. The filaments are also observable near the outer periphery of the Ce cell (some denoted by small white arrows in Fig. 4.13B), which may not be included in the keratinization of the Ce. As already mentioned, keratinization of the Ce is suggested to start from the formation of the exocuticle and its extension. The relationship between the Ci and Ce is delicate. Eventually they must be separated, but until that time of parting they probably remain tightly connected to ensure both the exchange of informational factors and the retention of synchronicity in the growth of the shaft and sheath.

Two Ci cells are shown in Fig. 4.14A. The inner side of the right cell (its nucleus denoted by a black arrow) adjoins the outer side of the left cell (nucleus denoted by a white arrow). The inner side-dominant keratinization of the right Ci cell is notable (Fig. 4.14A). Then the keratinization area is likely to expand to the whole cytoplasm as the left cell. Inner side-dominant keratinization is also clearly demonstrated in Fig. 4.14D, where the fibers are shown to be formed only on the right side of the degenerating nucleus (denoted by a thick arrow).

Final Steps in Cornification

In Fig. 4.15, the cornification of all cornifiable tissues has been largely accomplished. The junction between the Ci and Ce has become distinctly loose, in contrast to other cell-to-cell associations, including cornified cell-to-cornified cell junctions, although it is routinely unclear whether the actual split as shown in Fig. 4.15 occurs naturally or artificially. Movement of the split hair shaft during sectioning is one of the most difficult problems in studying the morphology of hair. The separation of the Ci and Ce likely involves a special mechanism, because in no other case are junctions between cornified tissues destroyed through physiological processes (Fig. 4.15).

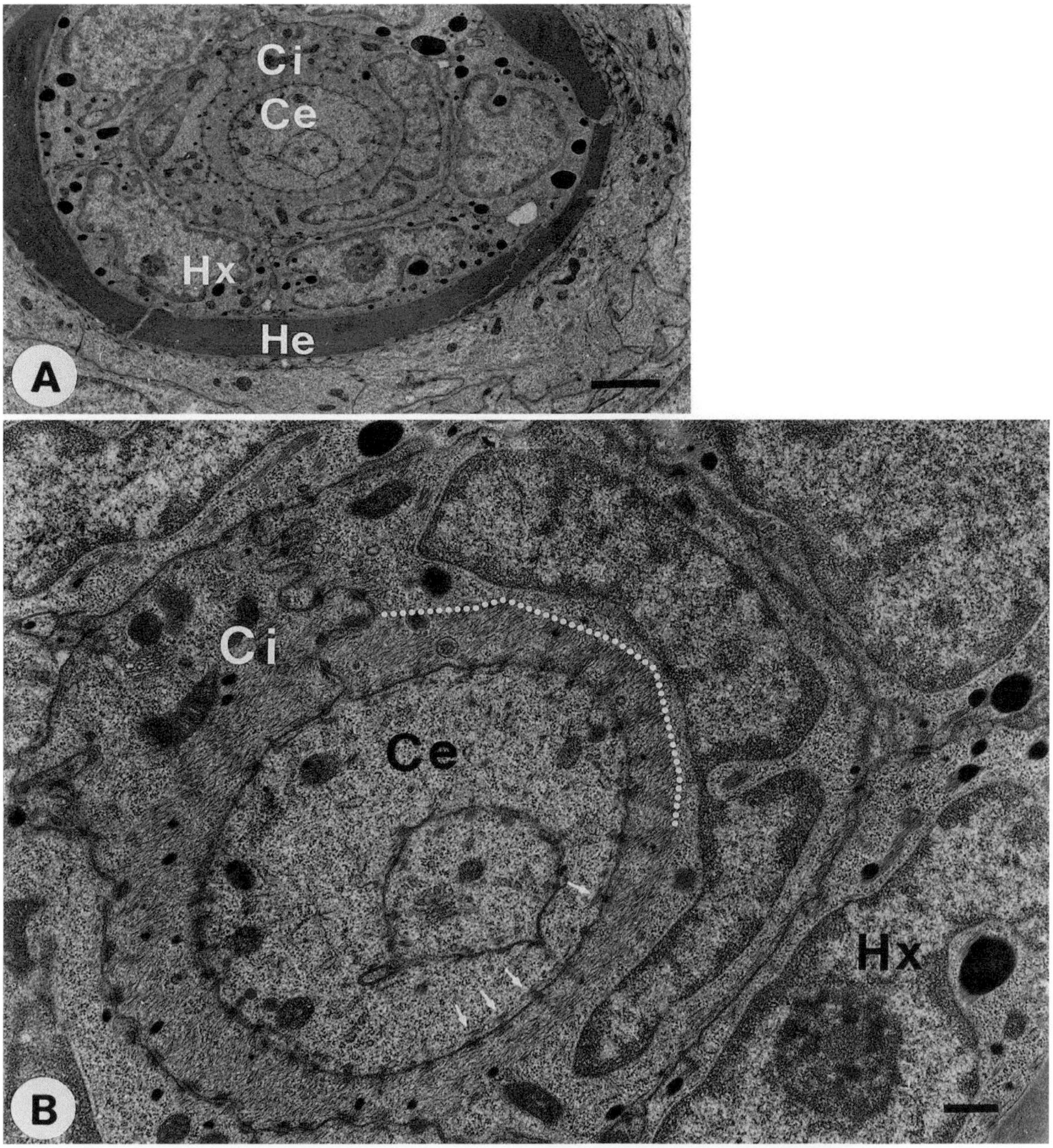

Fig. 4.13. Desmosomal junctions between Ci and Ce in an early stage. *He*, Henle's layer; *Hx*, Huxley's layer; *Ci*, cuticle of inner root sheath; *Ce*, hair cuticle. The meaning of the *white dotted line* in **B** is the same as that in Fig. 4.12, although it is not certain that those white dotted lines have the same physiological meaning. There are many desmosomes and associated filaments inside the line. White arrows in B denote tonofilaments near the cellular periphery. Dorsal skin of a day 2 rat was used. *Bars* **A** 2 µm; **B** 500 nm

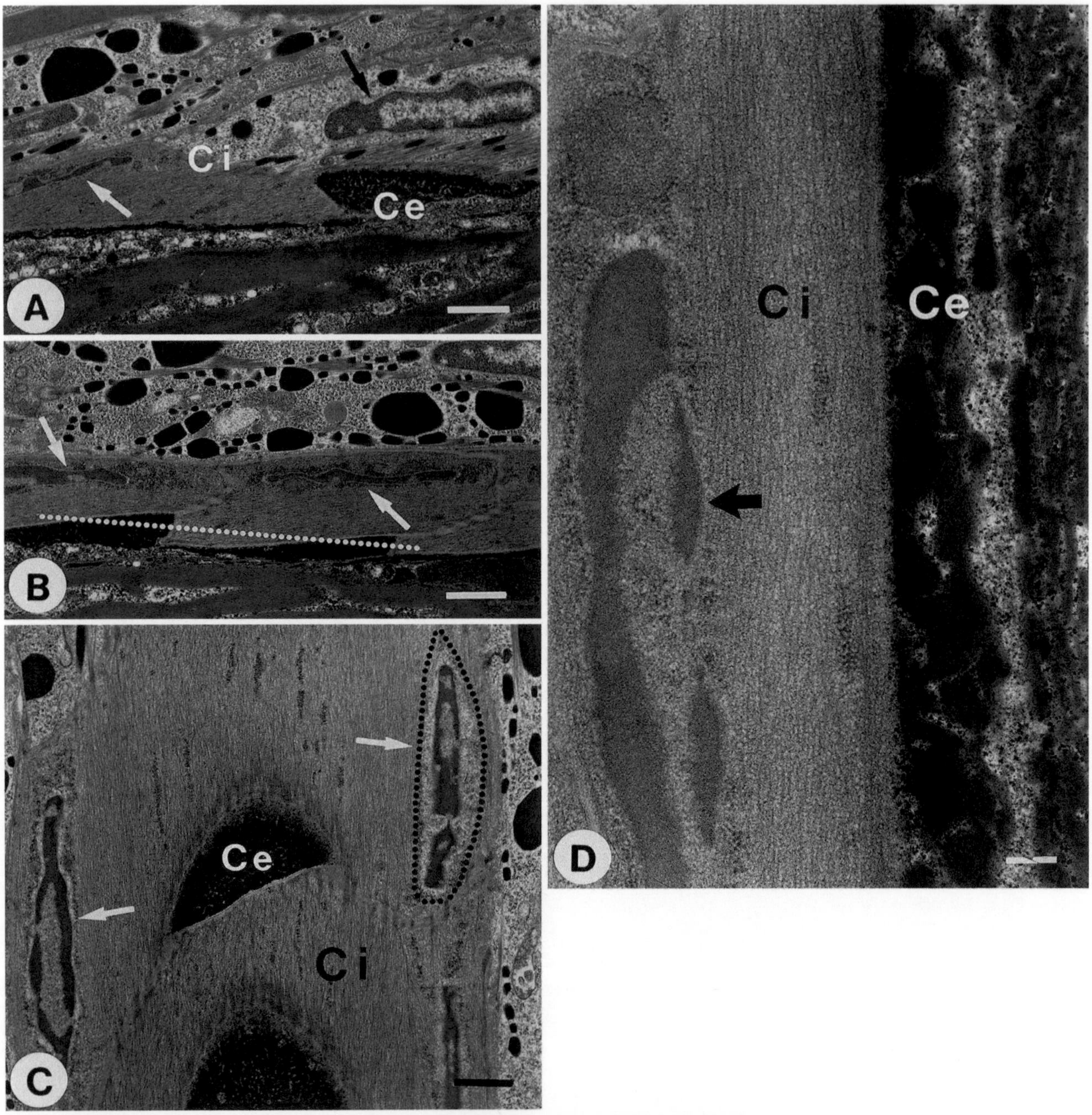

Fig. 4.14. Progress of Ci cornification and degeneration of the nuclei. *Ci*, cuticle of inner root sheath; *Ce*, hair cuticle. The *black arrow* in **A** denotes a nucleus of Ci before entering into the degrading process. The *white arrows* denote the disintegrating nuclei. The *thick arrow* in **D** represents a degraded nucleus of Ci. The *dotted line* in **B** represents the sectioning line of **C**. The nuclear area surrounded by a *dotted line* in **C** contains no fibers. **A**, **B**, **C** Dorsal skin of day 14 rats. **D** Dorsal skin of day 7 rat. *Bars* **A**, **B**, **C** 1 μm; **D** 200 nm

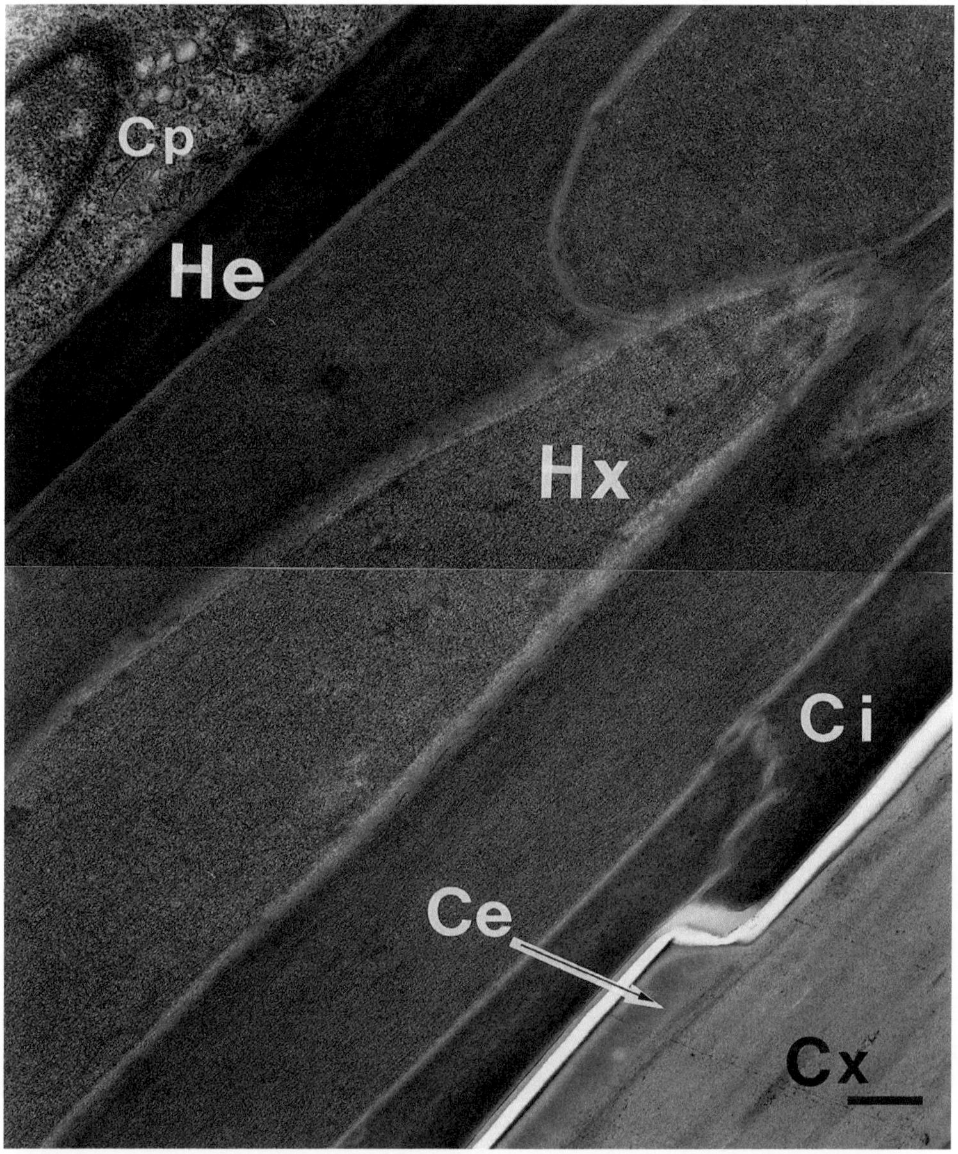

Fig. 4.15. Final stage of the cornification of hair tissues. *Cp*, companion layer; *He*, Henle's layer; *Hx*, Huxley's layer; *Ci*, cuticle of inner root sheath; *Ce*, hair cuticle; *Cx*, hair cortex. Sample was obtained from the dorsal skin of a day 7 rat. *Bar* 500 nm

Alternative Ways of Hair Emergence

Hair cuticle cells surrounding the hair cortex seem to occur in one of two patterns of arrangement, types A and B (Fig. 4.16). It is notable that all mammals identified to date show the type A pattern. If it is difficult to precisely adjust the speed of growth of the internal root sheath (IRS) and hair shaft, the growth of the shaft must be slightly faster than that of the IRS in the case of type A. The reverse case would result in the early separation of the IRS and shaft, because the shaft would be left behind. On the contrary, in type B, the growth of the IRS must be faster. A defect of type B is that the faster growth of the IRS hampers the process of desquamation. In other words, the IRS (Ci) sticks to the Ce as a result of the pressure exerted to raise them. This sticking in turn results in the encroachment of the Ce into the Ci (Fig. 4.16B; black areas in type B). In summary, the rapid growth of the Ci in type B disturbs the separation of the IRS from the shaft. Of course, rapid growth of the shaft causes early separation in type B. In contrast, more rapid elongation of the shaft would assist the extrusion of the IRS in type A. The ease of separation is also related to the relative hardness of the tissues concerned. In all likelihood, the adherence to the epidermal cornified layer (see Fig. 4.1) and the flexibility of the cell shape of the cuticle of the inner root sheath allow the desquamation of the IRS. If the IRS were harder than the shaft, desquamation would be a difficult process indeed.

When an animal needs to accelerate elongation of the hair shaft in response to the onset of cold weather, the IRS would be left behind in the case of type B, which would result in the formation of a deep crack around the shaft and an attendant risk of infection. In contrast, such cracking would not easily occur with the type A structure, which allows for extension of the IRS. Because the problem of heat can be met through epilation, it is more important that the hair prepare for cold. The type A structure is also advantageous with regard to the retention of sebum on the hair surface, which waterproofs the hair and prevents its snarling. Further study is needed to verify these hypothetical considerations.

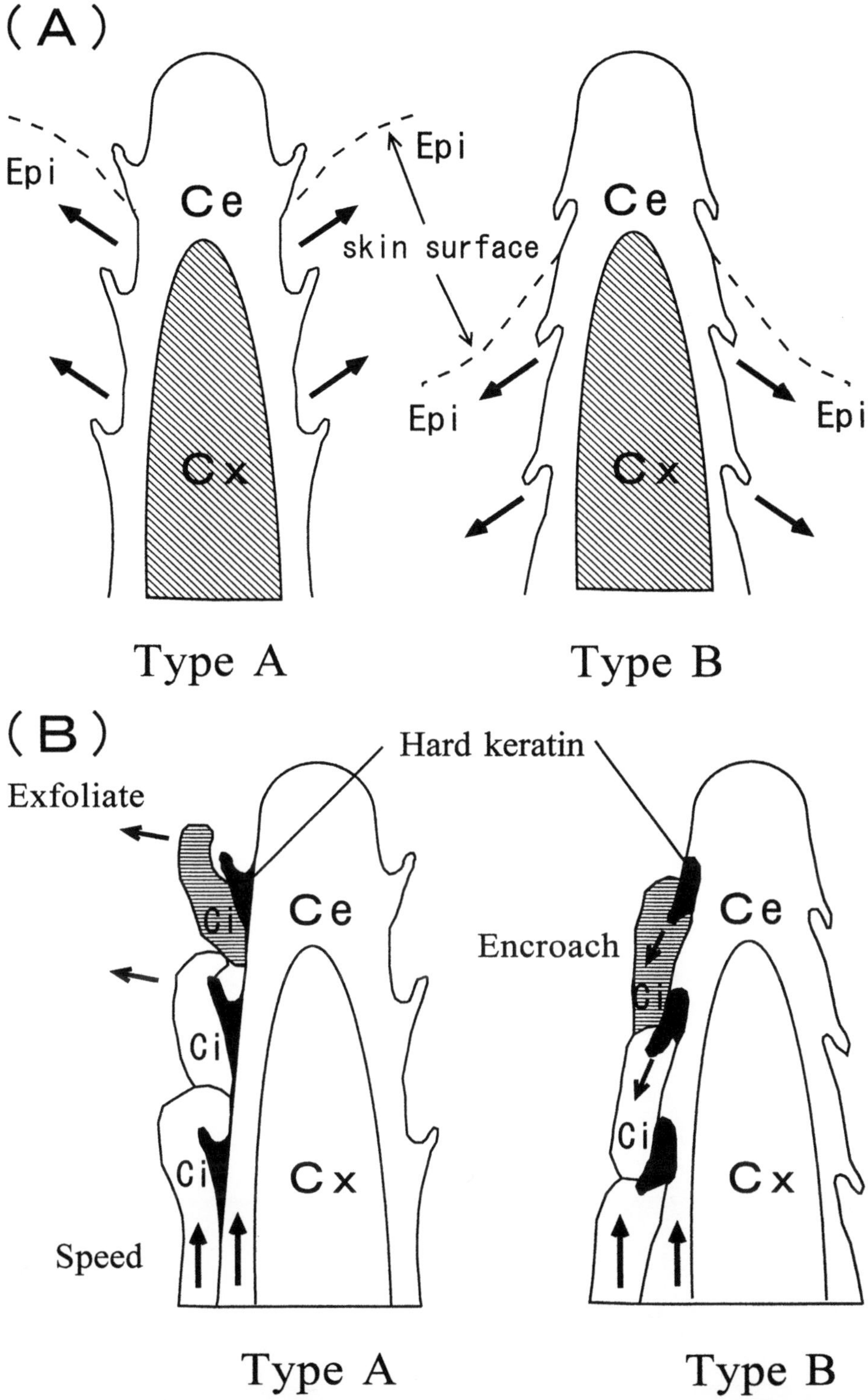

Fig. 4.16. Two ways of hair emergence depending on the arrangement of hair cuticle cells. **A** Morphology or ways of compilation of hair cuticle cells. **B** Effect of the cornification of hair cuticle cells on the fate of cuticle of inner root sheath cells. *Ci*, cuticle of inner root sheath; *Ce*, hair cuticle; *Cx*, hair cortex; *Epi*, epidermis; *broken lines*, skin surface. *Type A*, upper ends of the hair cuticle cells protrude from the hair shaft (real hair); *Type B*, lower ends of the hair cuticle cells protrude from the shaft (imaginary hair). *Arrows* designate the direction of the protrusion, and they also show the exfoliation of Ci cells in type A in part **B**

References

1. Birbeck MSC, Mercer EH (1956) The electron microscopy of the human hair follicle. Part 3. The inner root sheath and trichohyalin. J Biophys Biochem Cytol 3:223–237
2. Auber L (1950–1951) The anatomy of follicles producing wool-fibers, with special reference to keratinization. Trans R Soc Edinb LXII(part I)
3. Rochat A, Kobayashi K, Barrandon Y (1994) Location of stem cells of human hair follicles by clonal analysis. Cell 76:1063–1073
4. Taylor G, Lehrer MS, Jensen PJ, Sun TT, Lavker RM (2000) Involvement of follicular stem cells in forming not only the follicle but also the epidermis. Cell 102:451–461
5. Oshima H, Rochat A, Kedzia C, Kobayashi K, Barrandon Y (2001) Morphogenesis and renewal of hair follicles from adult multipotent stem cells. Cell 104:233–245
6. Panteleyev AA, Jahoda CA, Christiano AM (2001) Hair follicle predetermination. J Cell Sci 114:3419–3431
7. Orwin DFG (1976) Acid phosphatase distribution in the wool follicle. II. Henle's layer and outer root sheath. J Ultrastruct Res 55:325–334
8. Rothnagel JA, Rogers GE (1986) Trichohyalin, an intermediate filament-associated protein of the hair follicle. J Cell Biol 102:1419–1429
9. Lee S-C, Kim I-G, Marekov LN, O'Keefe EJ, Parry DAD, Steinert PM (1993) The structure of human trichohyalin. J Biol Chem 268:12164–12176
10. Powell BC, Rogers GE (1997) The role of keratin proteins and their genes in the growth, structure and properties of hair. In: Jolles P, Zahn H, Hocker H (eds) Formation and structure of human hair. Birkhauser, Basel
11. Parakkal PK, Matoltsy AG (1964) A study of the differentiation products of the hair follicle cells with the electron microscope. J Invest Dermatol 43:23–34
12. Rogers GE (1964) Structural and biochemical features of the hair follicle. In: Montagna W, Lobtiz W (eds) The epidermis. Academic, New York, pp 179–236
13. O'Guin WM, Sun T-T, Manabe M (1992) Interaction of trichohyalin with intermediate filaments: three immunologically defined stages of trichohyalin maturation. J Invest Dermatol 98:24–32
14. Tsukita S, Furuse M (1999) Occludin and claudins in tight-junction strands: leading or supporting players? Trends Cell Biol 9:268–273
15. Roth SI, Helwig EB (1964) The cytology of the dermal papilla, the bulb, and the root sheaths of the mouse hair. J Ultrastruct Res 11:33–51

5 Outer Root Sheath and Companion Layer

Conundrums

The outer root sheath (ORS) and companion layer (Cp) remain as mysterious tissues in the hair follicle, because they do not seem to undergo prominent cornification or any other terminal differentiation exhibiting distinct morphological specialization. The ORS is consecutive with the basal layer of the epidermis and is not involved in the production of either the hair shaft or skin keratinocytes. The general physiological roles of the ORS and Cp are not well known, but some investigators have revealed the presence of a hiding place for multipotent stem cells in the bulge area (BG) of the ORS [1, 2]. The position of this bulge area in the hair follicle is shown in Fig. 5.1. Stem cells in the bulge have been shown to migrate downward through the ORS at a speed of 70–100 μm per day with the mission to form cells of the internal root sheath and hair shaft by proliferating in the hair bulb [3]. Bulge cells were also reported to participate in the closure of an epidermal wound; in this case the cells migrated upward toward the skin surface [2, 3]. Thus, the ORS is at minimum constituted of native resident ORS cells and passenger stem cells traveling upward or downward through the ORS. In addition, the ORS is the site of attachment of the arrector pili (the smooth muscle that causes the hair to stand). The sebaceous gland is an appendage of the ORS. Note that all these findings concern the established hair follicle of adult animals, which is well enough known, whereas the cells in the developing ORS of neonatal animals are poorly understood. Stem cells in the neonatal ORS may migrate upward from the bulb to the bulge area seeking their niche.

The layer located between the ORS and He is the companion layer. According to the description of Orwin [4], this layer was discovered by Pinkus in 1927. Unfortunately, I have been unable to track down this paper. Orwin named it the "companion cell layer" because it was associated with He rather than the ORS [4]. The Cp cells were shown to move up at the same rate as the adjacent IRS cells [5]. In contrast, the resident ORS cells do not migrate or differentiate in the upward direction, unlike other follicle cells including those of the hair shaft, IRS, and Cp.

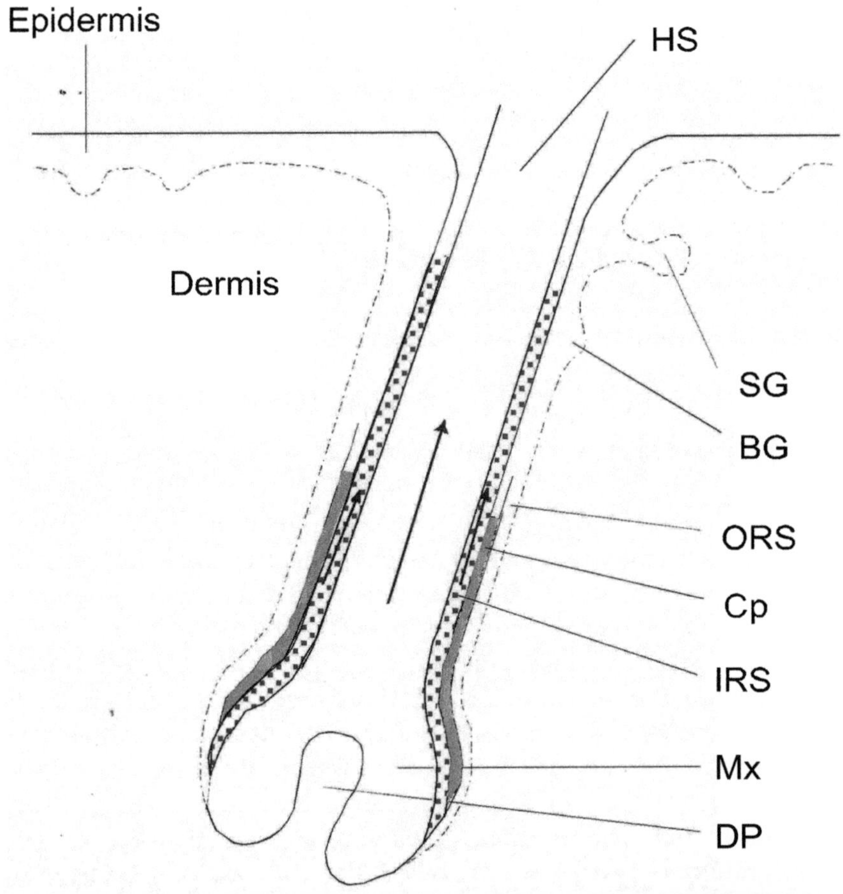

Fig. 5.1. Structure of a developed anagen hair. *HS*, hair shaft; *SG*, sebaceous gland; *BG*, bulge area; *ORS*, outer root sheath; *Cp*, companion layer; *IRS*, inner root sheath; *DP*, dermal papilla; *Mx*, hair matrix. *Gridded areas* represent the inner root sheath. The companion layer is colored gray based on the reported expression pattern of the keratin K6hf gene [3]. Although the mRNA of this keratin has recently been shown to appear also in upper hair matrix and medulla [10], it is still a helpful marker of Cp. *Broken lines* represent the basement membrane between dermal and epidermal tissues; *arrows* show the direction of cell movement

Outer Root Sheath in the Hair Bulb

At the lower area of the hair bulb, the outer root sheath (ORS) appears to be single layered (Fig. 5.2A). Such a structure means that the ORS touches Henle's layer (He) without the intervention of the companion layer (Cp). The sites of contact are designated by the pairs of tiny arrows facing each other in Fig. 5.2A. Trichohyalin granules, which serve as markers of the He layer at the height of the hair bulb, are observed in the relevant He. This finding suggests that the early differentiation of He is not induced by Cp, but is possibly dependent on the ORS. In contrast, at the upper area, the ORS becomes double layered, with one layer probably being the ORS and the second corresponding to the Cp (Fig. 5.2B). Both granules and tonofilaments are seen in He at the height of Fig. 5.2B. The next inner layer of the ORS in Fig. 5.2B is thought to be the Cp on the basis of the special morphology of the boundary between the ORS and early Cp, to be described later. The thickness of the ORS is 0.2 μm or less in some areas, suggesting the presence of increasing pressure from inside to outside at this height of the follicle.

Identification of the Companion Layer

The morphological characteristics of the innermost layer of the ORS have been reported by Ito [6, 7]. Winter and colleagues have reported the specific expression of K6hf, one of the type II cytokeratins in the companion layer (Cp) of human hair [8]. The gray areas in Fig. 5.1 are drawn based on their study of K6hf mRNA expression. The gray area (cylindrical in an actual hair) extends to about twice the height of the hair bulb, but it is of course not identical with the whole area of the Cp. The K17 messenger is reported to be present at the upper Cp, at which the mRNA of K6hf is not expressed, but unfortunately K17 is not specific to the Cp [9]. Furthermore, recent research revealed that K6hf is expressed also in the upper hair matrix and medulla [10], showing that it is still imperfect to identify the companion layer.

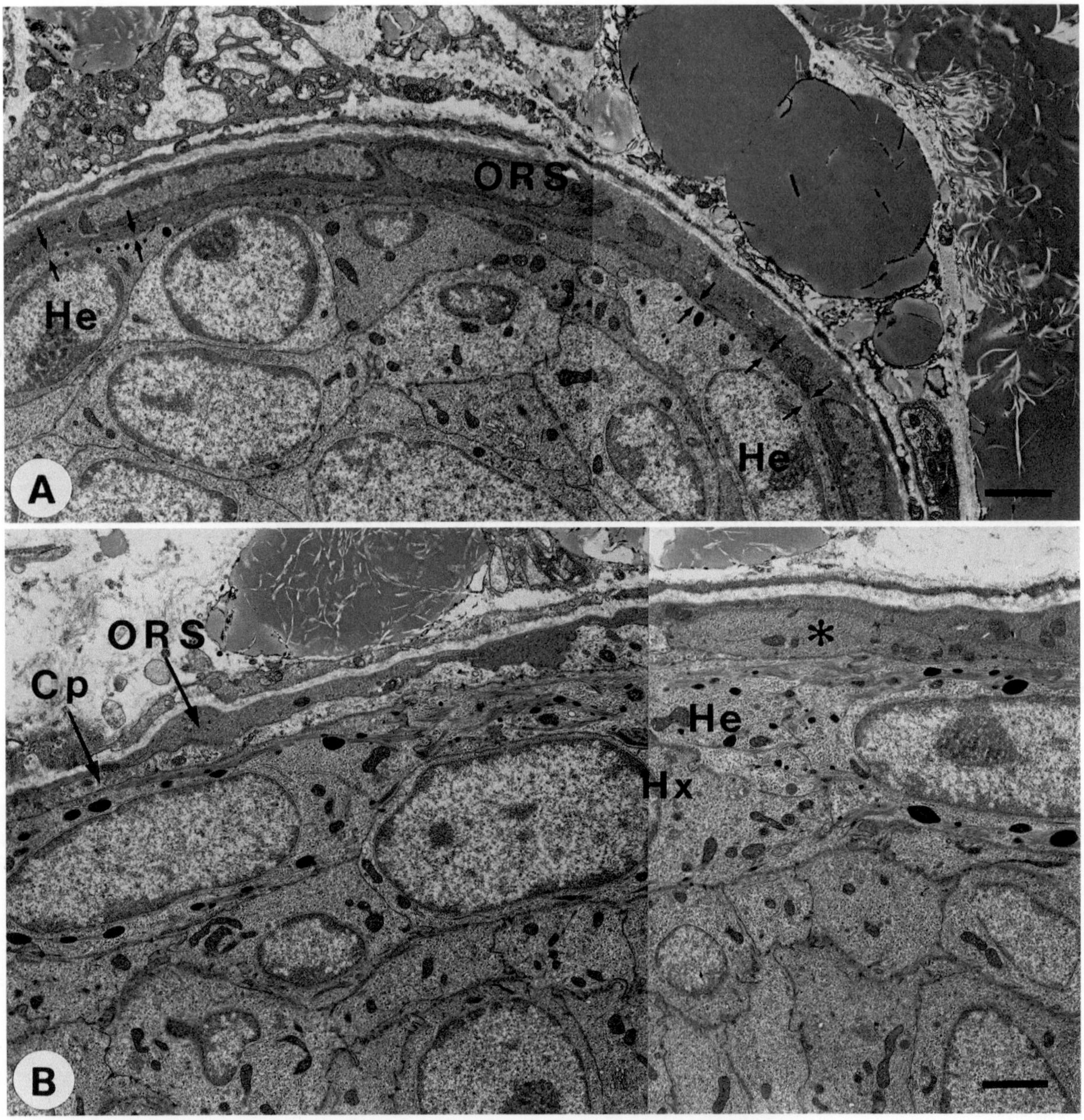

Fig. 5.2. Outer root sheath and companion layer at the level of hair bulb. *ORS*, outer root sheath; *Cp*, companion layer; *He*, Henle's layer; *Hx*, Huxley's layer. The *arrows* in **A** show direct contact areas of ORS and He without the intervention of Cp at the lower bulb region. **B** The intervention of the Cp at the upper bulb region. Cell denoted by the *asterisk* was not identified whether it is affiliated to ORS or to Cp. Samples were obtained from the dorsal skin of rats on day 7 after birth. *Bars* 2 μm

Boundary and Overlapping of the Outer Root Sheath and Companion Layer

Above the hair bulb, both the outer root sheath (ORS) and companion layer (Cp) often occur as multilayers (Fig. 5.3A,C) [11]. The structure of the Cp–Cp and Cp–ORS boundaries is impressive (Fig. 5.3A–C; also shown schematically in Fig. 5.4). The boundaries (thick arrows in Fig. 5.3) have many inflection points, each of which is often accompanied by a desmosome (arrowheads in Fig. 5.3A, an arrow in Fig.5.3B, asterisks in Fig. 5.4; see also Fig. 4.6). The desmosomes are often accompanied by the tonofilaments (Figs. 5.3, 5.4). These special structures are not found at the boundaries of any other hair follicle cells, including those at the Cp–He boundary. The number of overlapping cell layers is variable: for example, one layer of ORS is accompanied by four layers of Cp in Fig. 5.3A, whereas three layers of ORS accompany two layers of Cp at the position of the dotted line in Fig. 5.3C. These findings suggest that previous observations that the Cp is a single layer (6, 7) may not be universal. Figure 5.3C is a rare case in which the positions of the ORS and Cp seem to be partially inverted. The asterisk denotes the ORS cell and the circle the Cp cell, as identified by the structures of the cellular boundaries (thick arrows in Fig. 5.3C). The uppermost Cp cell seems to be sandwiched between ORS cells.

Thick hair does not necessarily entail a thick ORS–Cp. A section of human scalp hair from a subject aged 22 years in the textbook of Bloom and Fawcett shows a thin outer sheath consisting of only an ORS of one layer and a Cp of one layer where the cortex consists of five to six -layers at the level of the upper part of the hair bulb [12]. Of course, identification of the ORS and Cp by electron microscopy is not perfect. The number of ORS layers usually increases as it grows from the lower to the middle part of the hair follicle. Another feature of the Cp cell is its flat shape, in contrast to the columnar shape of the ORS cell (Fig. 5.3A). At this height, in which both He and Hx are fully keratinized, the difference in electron density of the cytoplasm of the ORS and Cp may become ambiguous, as in Fig. 5.3C.

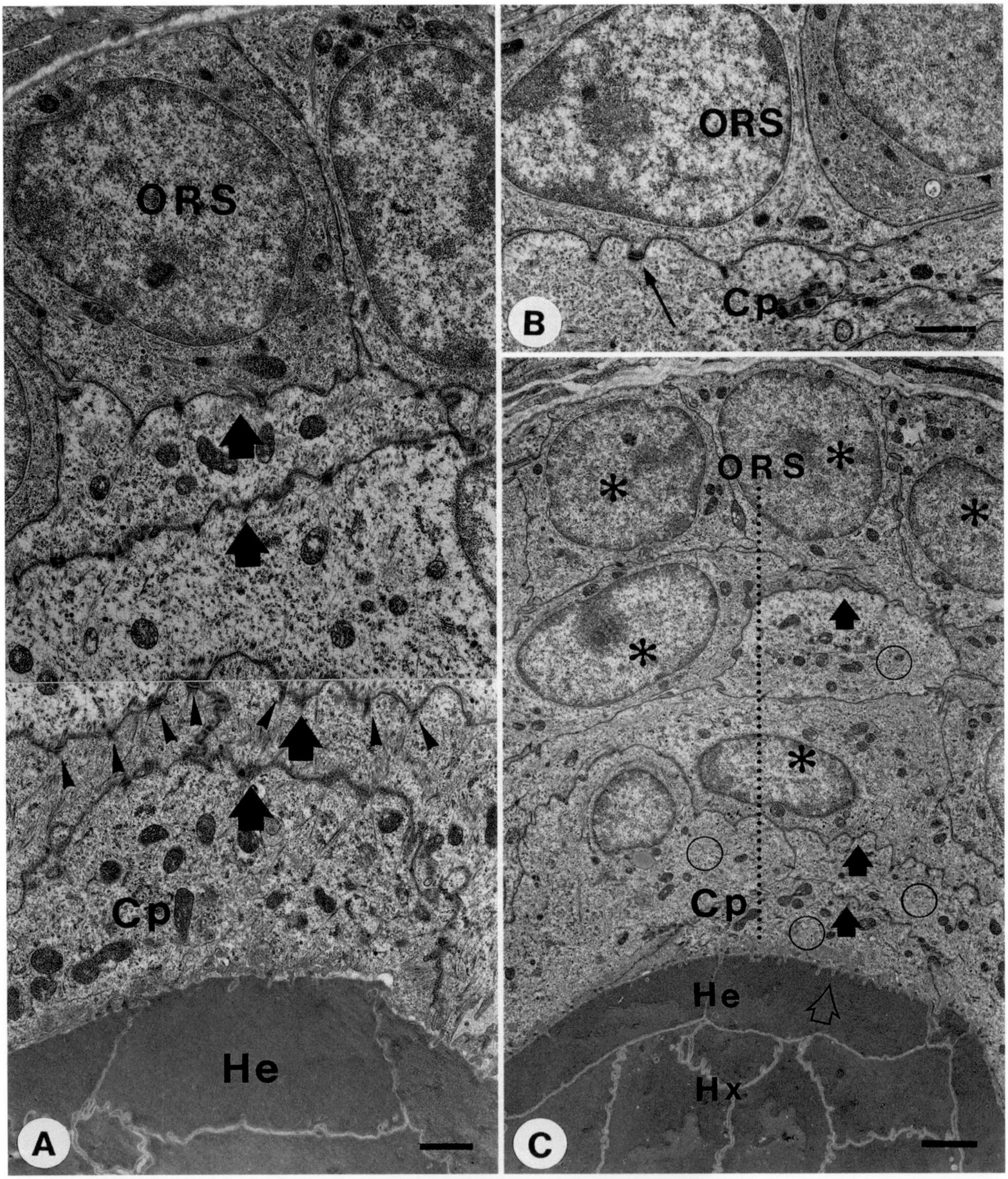

Fig. 5.3. Specific structures of the boundary between outer root sheath and companion layer and irregular pile of the cellular layers. *ORS*, outer root sheath; *Cp*, companion layer; *He*, Henle's layer; *Hx*, Huxley's layer; *thick arrows*, distinctive structure of the boundary between ORS and Cp. Both *arrowheads* in **A** and a *thin arrow* in **B** denote inflection points of the cell membrane contours accompanied by the desmosomes. *Asterisks*, ORS cells; *circles*, Cp cells. *Dotted line* in **C** was drawn to explain the order and overlapping of the cellular layers. The *open arrow* in **C** shows an active feature of the membrane of a He cell. Samples in **A** and **C** were obtained from the facial skin of day 2 rats; that in **B** was from the dorsal skin of a rat at day 2 after birth. *Bars* **A**, **B** 1 μm; **C** 2 μm

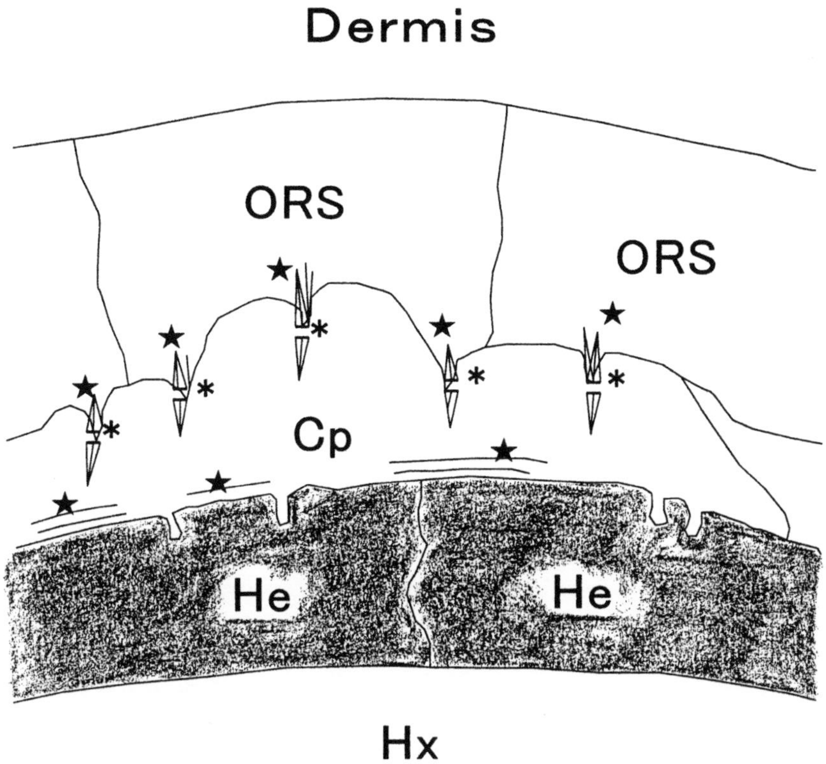

Fig. 5.4. A schematic representation of the outer side of the hair follicle. *ORS*, outer root sheath; *Cp*, companion layer; *He*, Henle's layer; *Hx*, Huxley's layer. The boundaries between ORS and Cp include "mountains" and "ravines." In most cases a desmosome (*) is found at the bottom of the ravine. The desmosomes are usually accompanied by tonofilaments (★) in the cytoplasm of both ORS and Cp cells. The tonofilaments are also found frequently in the near-He area of the cytoplasm of Cp cells

Tonofilaments in the Companion Layer

Tonofilaments often appear in the vicinity of He of Cp cells (arrows in Fig. 5.5A) [6, 7]. The enlarged view in Fig. 5.5B shows the partial keratinization of a Cp cell (asterisk in Fig. 5.5B). The fibers appeared to be wavy in comparison to those in the cells of other layers and are arranged horizontal to the He surface, with a 0.4- to 0.5-µm width in Fig. 5.5B. In contrast to the fibers adherent to the desmosomes in the opposite (ORS) side, the He-side fibers seem to be independent of the desmosomes. The Cp may communicate with the keratinized He through the surviving mem-brane traffic system in He (for example, the open arrow in Fig. 5.3C suggests the endocytosis of keratinized He), which is also illustrated in Fig. 5.4. Although partial keratinization is evident as shown here, I have not encountered a Cp cell with a cytoplasm entirely filled with keratin fibers.

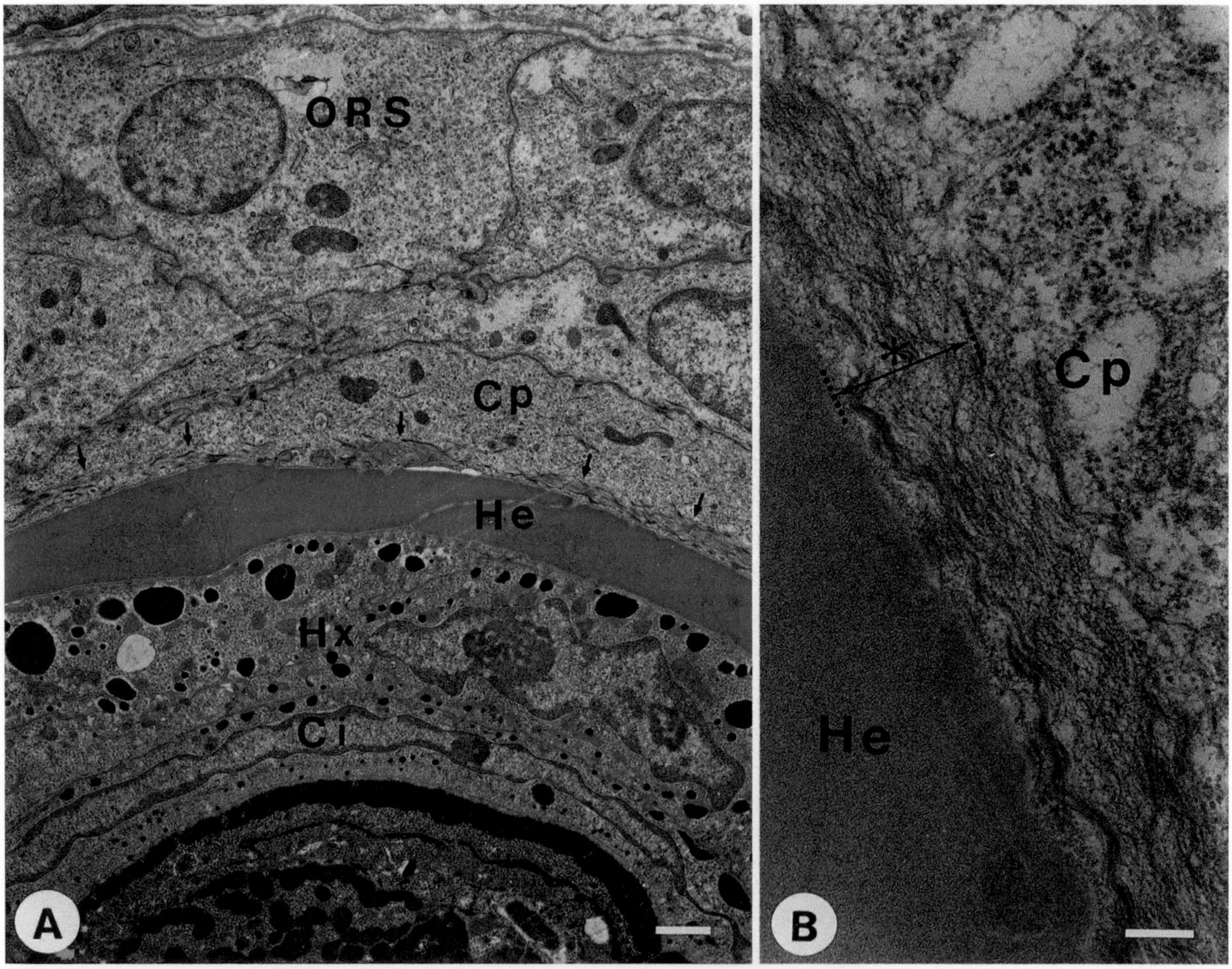

Fig. 5.5. Formation of keratin filament bundles in the companion layer. *ORS*, outer root sheath; *Cp*, companion layer; *He*, Henle's layer; *Hx*, Huxley's layer; *Ci*, cuticle of inner root sheath. *Arrows* in **A** and the *asterisk* in **B** denote the keratin filaments. The width of the keratinized area is shown to be 0.4–0.5 µm in this case (as shown by the *dotted lines* and *double-headed arrow* in **B**). Samples were obtained from the dorsal skin of day 2 rats. *Bars* **A** 1 µm; **B** 200 nm

General Characteristics of the Outer Root Sheath

Outer root sheath (ORS) cells undergo cell division at the level of the hair bulb (Fig. 5.6A), as well as at the upper level at which the inner root sheath is heavily keratinized (Fig. 5.6B). The structure of the boundary between adjacent ORS cells often becomes highly complicated, as the arrows in Fig. 5.7A,B show. These photographs were taken at levels higher than the bulb. In contrast to a simple line between the ORS and Cp, the ORS–ORS boundary is delineated by a complicated curvature with a wide gap (see arrow in Fig. 5.7A). A similar structure is shown in Fig. 5.7B, in which the cellular cytoplasm has darkened, probably as a result of ORS-specific cornification.

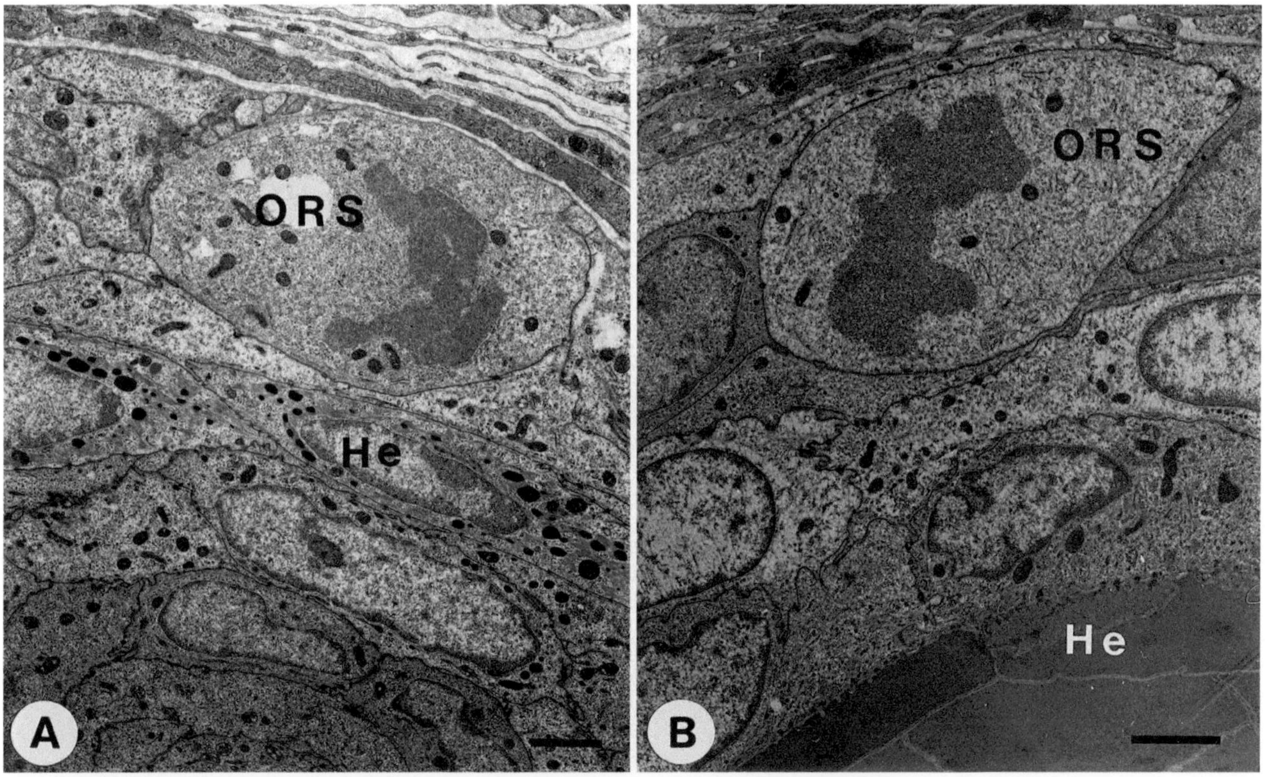

Fig. 5.6. M-phase cells of outer root sheath. *ORS*, outer root sheath; *He*, Henle's layer. Henle's layer is not cornified in **A**, whereas it has been cornified in **B**. Samples were obtained from the dorsal skin of day 2 rats. *Bars* 1 μm

Further complicated structures characteristic of some ORS cells are shown in Fig. 5.8. The arrows in Fig. 5.8A,B show that the structure of the cell boundary is markedly complicated, and it is difficult to distinguish the vacuoles, vesicles, and cytoplasm from each other. This type of ORS cell is commonly found in the area above the hair bulb. Some cells appear to contain the vacuoles (Fig. 5.8A), while others may contain translucent or coarse structural components (Figs. 5.7B, 5.8C–E), which are discussed next. Although lysosome-like structures are seen (arrows in Fig. 5.8D,E), they are not particularly abundant.

As shown in Fig. 5.8, the cytoplasm of ORS cells contains many electron-lucent spaces that are not encircled by membranous structures. These spaces are electron lucent but not vacant, in sharp contrast to the cytoplasm of Cp cells, which appears homogeneous and unremarkable (Fig. 5.8A,C). One can find many small vesicles in the outer side of ORS cells (arrowhead in Fig. 5.9). These immature (undifferentiated) ORS cells probably have highly active membrane traffic systems (Fig. 5.9). Hemidesmosomes, formed in the basal layer cells of the epidermis, are also seen in the ORS (arrows in Fig. 5.9), but I believe they are not as frequent as in the epidermis.

Fig. 5.7. Boundaries between outer root sheath cells and connective tissue follicle cells. *CTF*, connective tissue follicle; *ORS*, outer root sheath; *Cp*, companion layer. The *asterisks* denote the space that corresponds to the basement membrane between the skin epidermis and mesenchymal dermis. Dorsal skin from day 7 rats was used. *Bars* **A** 1 μm; **B** 500 nm; **C** 200 nm

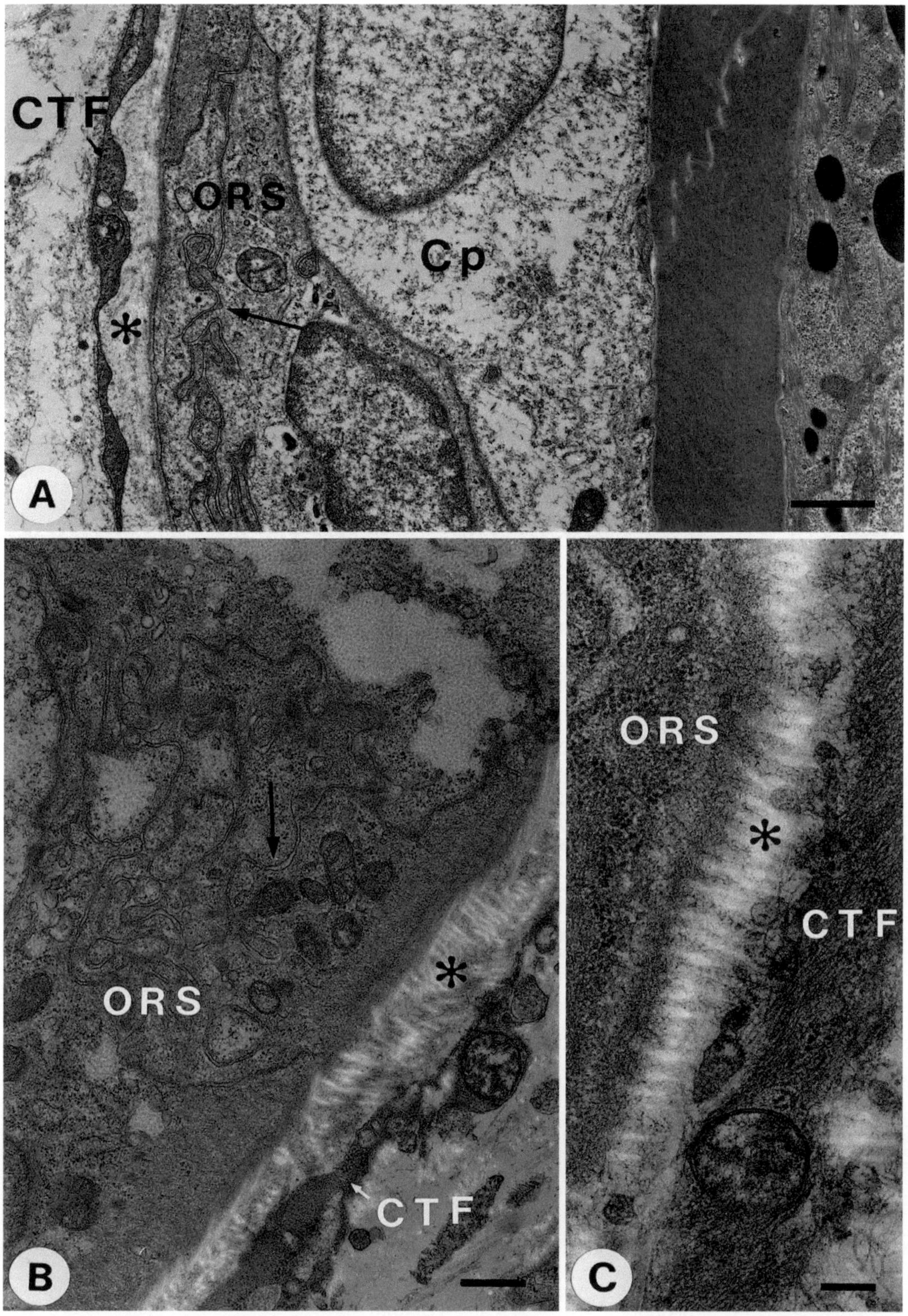

Fig. 5.7.

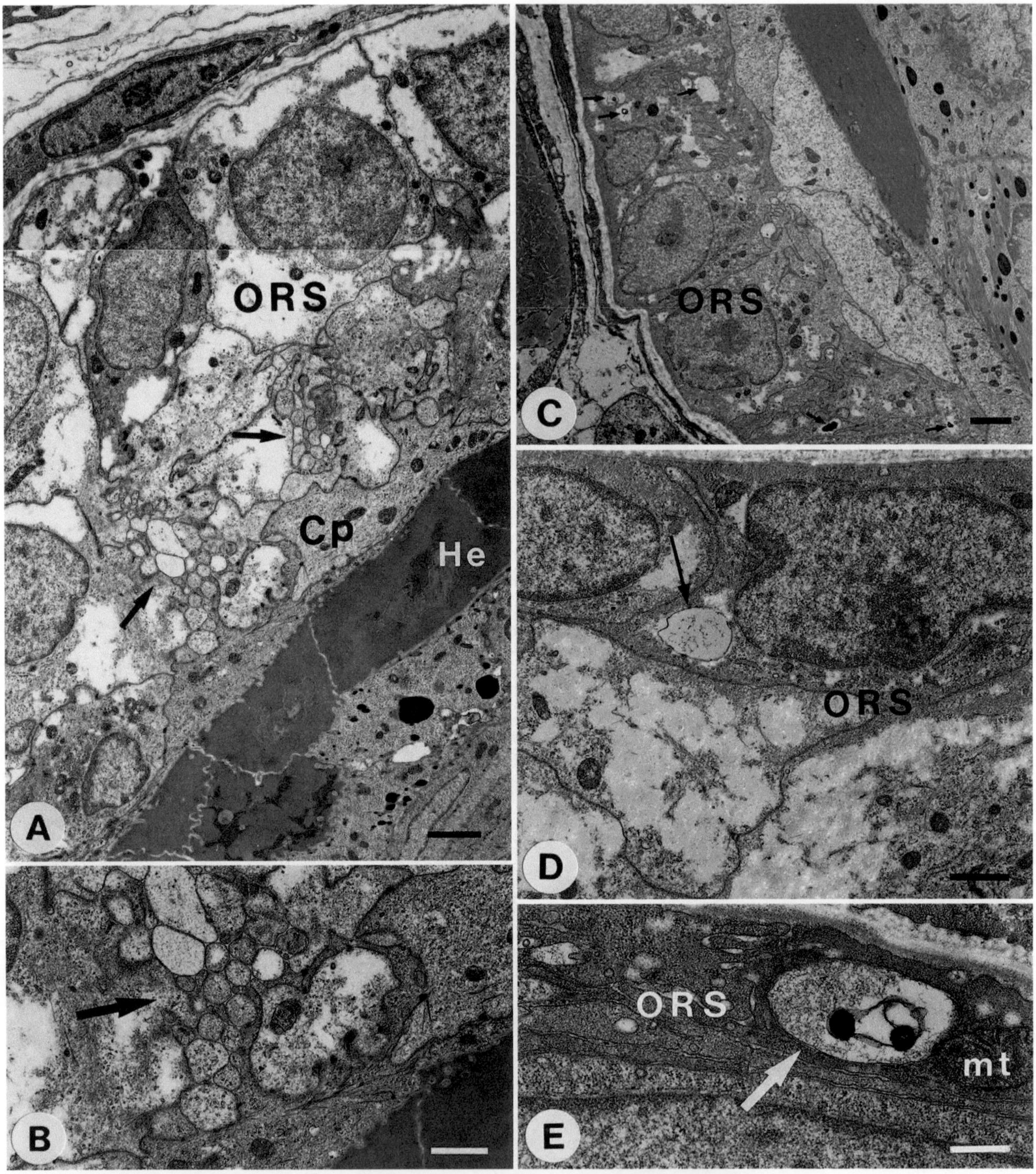

Fig. 5.8. Cytoplasmic structures in the cells of outer root sheath. *ORS*, outer root sheath; *Cp*, companion layer; *He*, Henle's layer; *mt*, mitochondria. *Arrows* in **A** and **B** denote the complicated structures of ORS cell–ORS cell boundaries; *arrows* in C denote the electron-lucent spaces; *arrows* in **D** and **E** denote the lysosome-like structures. Dorsal skins from day 2 (**A**, **B**), day 7 (**C**, **D**), and day 14 (**E**) rats were used for the photography. *Bars* **A** 2 μm; **B** 1 μm; **C** 2 μm; **D** 1 μm; **E** 500 nm

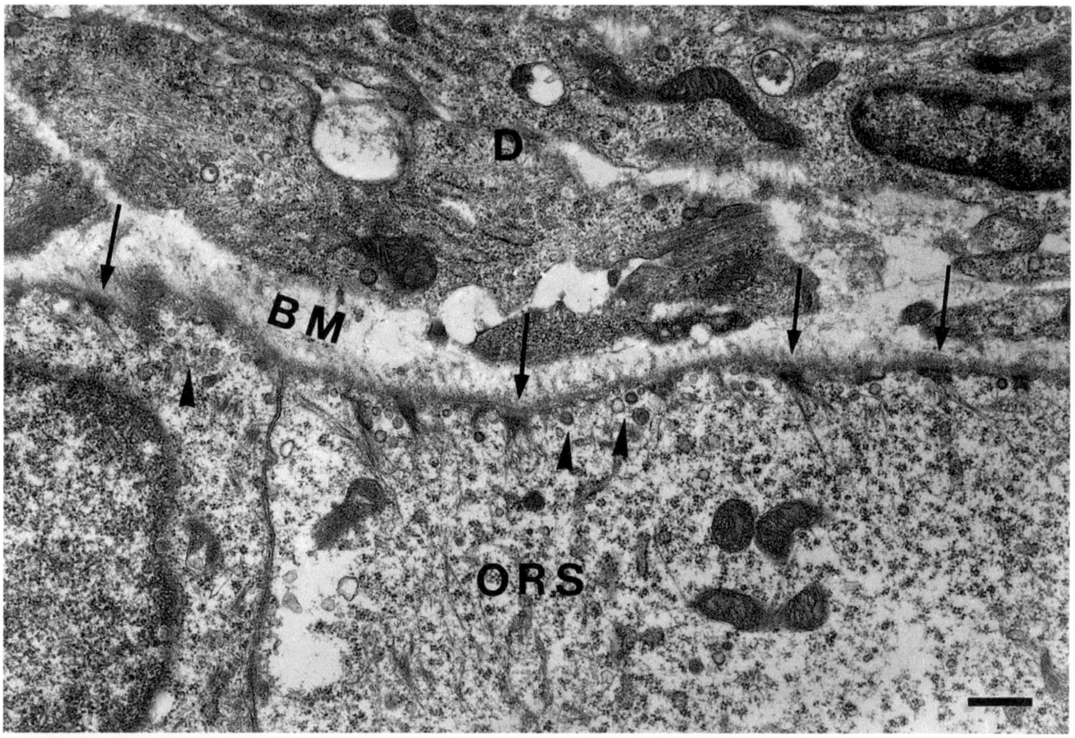

Fig. 5.9. Peripheral area of outer root sheath. *D*, dermis; *BM*, basement membrane; *ORS*, outer root sheath; *arrows*, hemidesmosome; *arrowheads*, vesicles. Dorsal skin from day 0 rats was used for the experiment. *Bar* 500 nm

Comparison with Fetal Epidermis

In contrast to epidermal spinous layer cells of postnatal animals, which have perinuclear electron-lucent areas (13), the ORS cell has a perinuclear electron-dense area, as shown in Fig. 5.8A,C. Intriguingly, this cytoplasmic pattern is similar to that of fetal skin. Figure 5.10 shows the dorsal skin of rats at day 18 of gestation (day 21 is usually the day of birth). The cells corresponding to the spinous layer of postnatal animals contain many vacuoles (some designated by the asterisk or a, b, c in

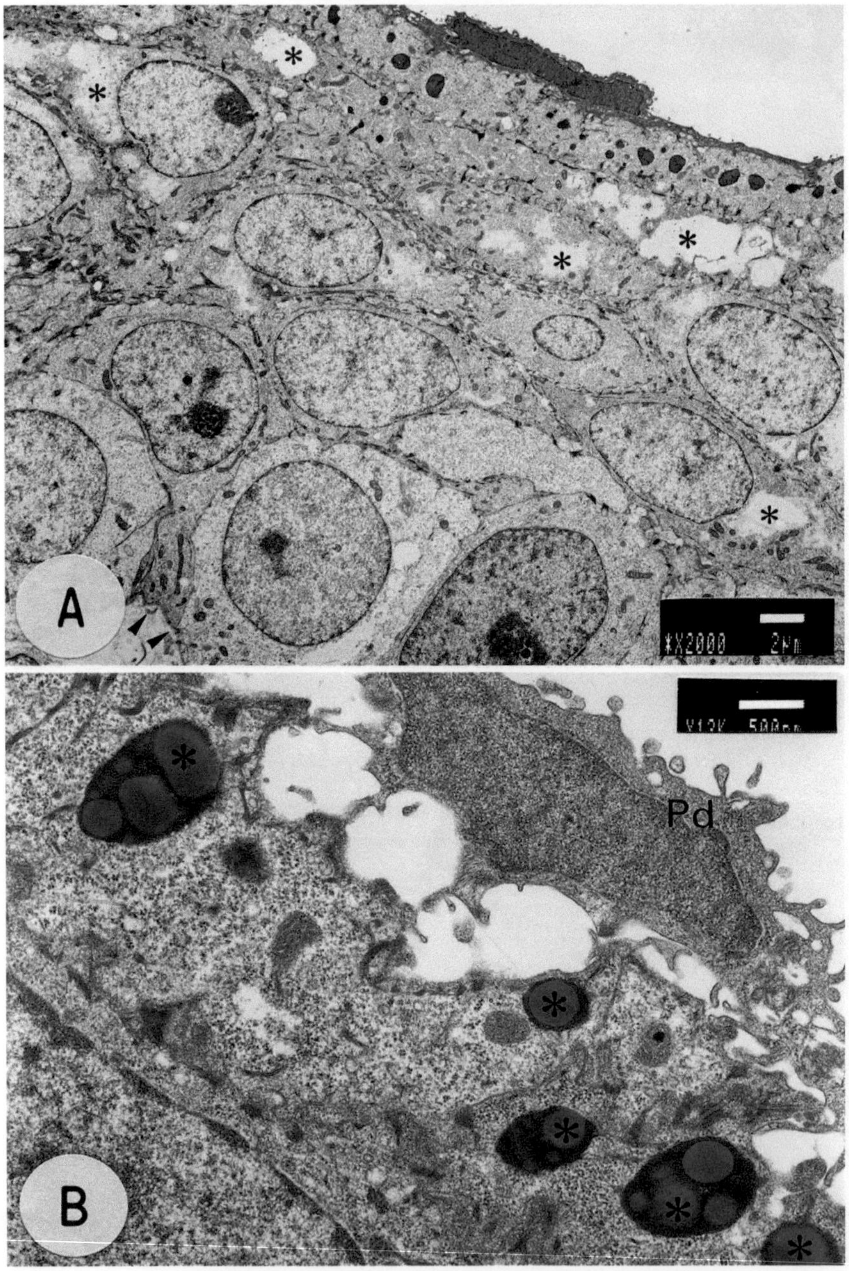

Fig. 5.10.

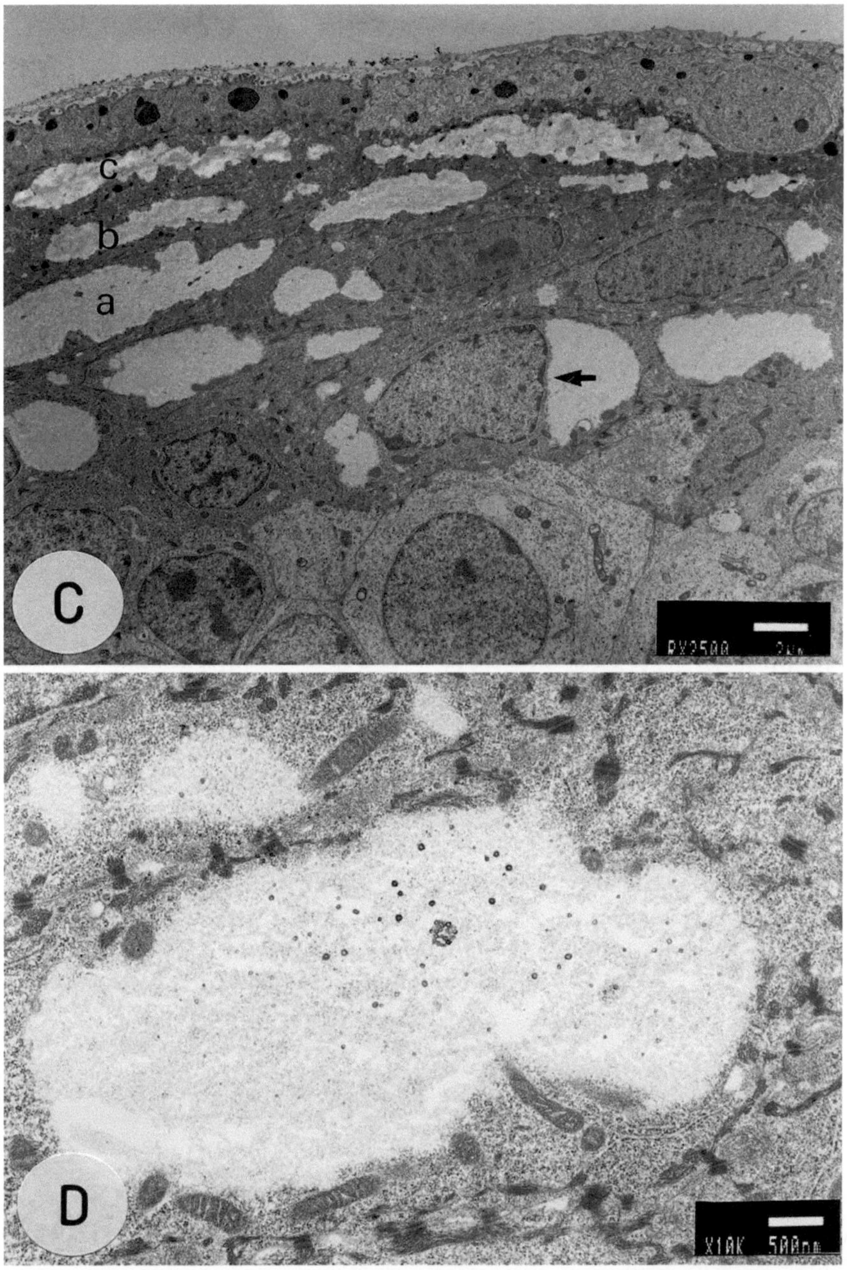

Fig. 5.10. Structure of skin of fetal rat. **A** Cells differentiate from lower stem cells to upper granule-rich cells. The *asterisks* denote the vacuoles in the intermediate layers. **B** Granules (*) in the uppermost layers. *Pd*, periderm. The granules are likely to not have keratohyalin or keratins. They may be composed of lipid-related constituents. **C** This photograph was prepared on soft printing paper (FUJIBRO WP FM2); it enables observance of the intravacuolar materials (*a*, *b*, *c*). The *arrow* denotes the perinuclear electron-dense area. **D** Enlarged view of a nearly vacant type of vacuole. Samples were obtained from the dorsal skin of rats at day 18 of gestation. *Bars* **A** 2 μm; **B** 500 nm; **C** 2 μm; **D** 500 nm

Fig. 5.10A–C), which are not vacant. Figure 5.8D shows that the vacuoles in ORS cells are also not vacant. The arrow in Fig. 5.10C denotes the perinuclear electron-dense area. The complicated cellular boundaries seen in the ORS (Fig. 5.8A,B) are not observed among fetal skin cells.

Final Fate

In contrast to the case of the hair shaft and internal root sheath, cells have been shown to be capable of migrating through the outer root sheath (ORS) [1–3]. I show in Fig. 5.6 the division of cells of the ORS at the lower level, in which He cells are just commencing differentiation (Fig. 5.6A), as well as at the upper level, in which the cytoplasm of the Hx cells is heavily keratinized (Fig. 5.6B). These findings suggest that new cells can be supplied at any time and place except in the lower half of the hair bulb. In comparison with the ordinary ORS cells in Fig. 5.11A, those in Fig. 5.11B have undergone ORS-specific cornification. Their cytoplasm is generally electron dense and contains few of the electron-lucent areas that are commonly seen in the ORS cells in Fig. 5.11A. The outer layer (ORS1) of the double-layered ORS in Fig. 5.11B seems to be on the way to separate from the hair follicle, implying that these cells desquamate into the dermal matrix. The ORS1 may correspond to the spinous layer of the epidermis, although they are not identical. It is notable that the structure of the basement membrane is altered, as can be seen by comparing Fig. 5.11A with Fig. 5.11B. I speculate that two or more overlapping ORS layers are needed to execute the desquamation of the ORS, because the loss of the single-layer ORS would result in the direct contact of the companion layer with the dermal matrix, which I have not encountered.

Fig. 5.11. Desquamation of outer root sheath cells. *CTF*, connective tissue follicle; *ORS*, outer root sheath; *Cp*, companion layer; *He*, Henle's layer. **A** Before desquamation. **B** Desquamating outer root sheath (*ORS 1*) cells. Samples were obtained from the dorsal skin of rats at day 7 after birth. *Bars* **A** 2 μm; **B** 500 nm

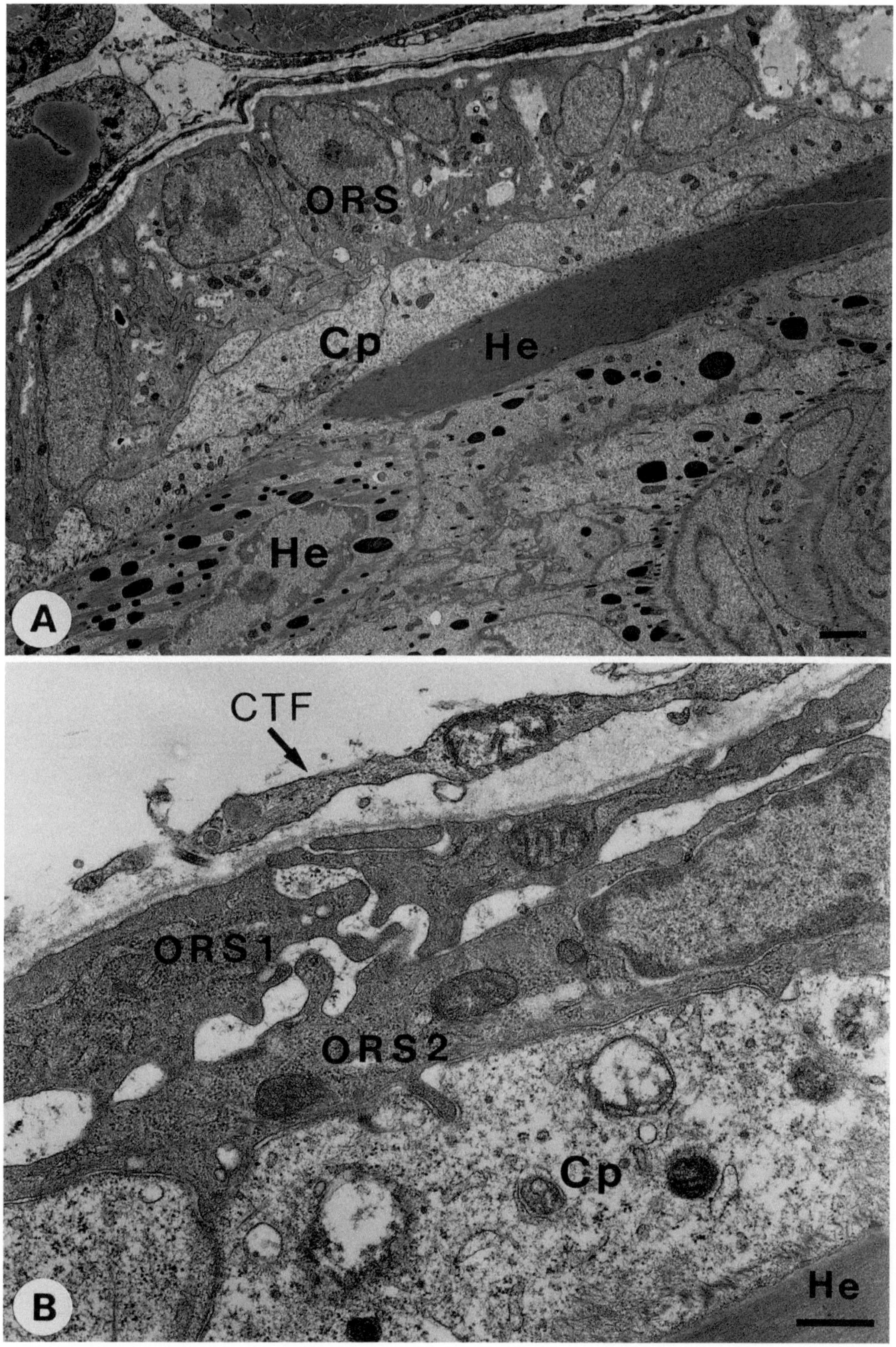

Fig. 5.11.

References

1. Cotsarelis G, Sun T-T, Lavker RM (1990) Label-retaining cell reside in the bulge of the pilosebaceous unit: implications for follicular stem cells, hair cycle, and skin carcinogenesis. Cell 61:1329–1337
2. Taylor G, Lehrer MS, Jensen PJ, Sun T-T, Lavker RM (2000) Involvement of follicular stem cells in forming not only the follicle but also the epidermis. Cell 102:451–461
3. Oshima H, Rochat A, Kedzia C, Kobayashi K, Barrandon Y (2001) Morphogenesis and renewal of hair follicles from adult multipotent stem cells. Cell 104:233–245
4. Orwin DF (1979) The cytology and cytochemistry of the wool follicle. Int Rev Cytol 60:331–374
5. Chapman RE (1971) Cell migration in wool follicles of sheep. J Cell Sci 9:791–803
6. Ito M (1986) The innermost cell layer of the outer root sheath in human anagen hair follicle. Light and electron microscopic study. Arch Dermatol Res 279:112–119
7. Ito M (1988) Electron microscopic study on cell differentiation in anagen hair follicles in mice. J Invest Dermatol 90:65–72
8. Winter H, Langbein L, Praetzel S, Jacobs M, Rogers RA, Leigh IM, Tidman N, Schweizer J (1998) A novel human type II cytokeratin, K6hf, specifically expressed in the companion layer of the hair follicle. J Invest Dermatol 111:955–962
9. McGowan KM, Coulombe PA (2000) Keratin 17 expression in the hard epithelial context of the hair and nail, and its relevance for the pachyonychia congenita phenotype. J Invest Dermatol 114:1101–1100
10. Wang Z, Wong P, Langbein L, Schweizer J, Coulombe PA (2003) Type II epithelial keratin 6hf (K6hf) is expressed in the companion layer, matrix, and medulla in anagen-stage hair follicles. J Invest Dermatol 121:1276–1282
11. Hashimoto K (1988) The structure of human hair. Clin Dermatol 6:7–21
12. Bloom W, Fawcett DW (1975) A textbook of histology, 3rd edn. Saunders, Philadelphia
13. Morioka K, Takano-Ohmuro H, Sameshima M, Ueno T, Kominami E, Sakuraba H, Ihara S (1999) Extinction of organelles in differentiating epidermis. Acta Histochem Cytochem 32:465–476

Hair Bulb and Papilla

6

Overview

The hair bulb consists of the immature outer and inner root sheaths, precursors of the hair shaft, hair matrix, dermal papilla, and surrounding connective tissue follicle. All but the dermal papilla have already been mentioned. In the adult hair, the dermal papillae often maintain the vascular system that is responsible for the nutritional support and hormonal regulation required for hair growth [1–3]. Because epidermal cells have no ability to maintain a vascular system in their environment, nutritional support from the dermis is thought to be important. In contrast to adult skin, the vascular system in neonatal skin is not yet well developed and angiogenesis in the papilla is rarely observed ([4, 5]; photographs in this chapter). The papilla is of course thought to provide not only nutrient substances, but also regulators of growth and differentiation of the matrix cells [6, 7]. Hormones such as androgen that regulate hair growth may also operate via papilla cells [6, 7].

Figure 6.1 illustrates the formation of the dermal papilla. In morphology it is largely spherical in the early stage (Fig. 6.1B), often becoming cylindrical (far from the shape of a "papilla") later (Fig. 6.1C). Its morphology also depends on the diameter of the hair shaft: thick hair is accompanied by a large spherical bulb with a large spherical papilla even in the later stages. The papilla is initially filled with dermal cells, which are often later gradually replaced by intercellular matrix (Fig. 6.1D).

Fig. 6.1. Outlines of development and maturation of dermal papilla. *DP*, dermal papilla; *Mx*, hair matrix. **A** Initial step of the intrusion of papilla into hair matrix cells. *Dotted line* denotes the boundary between the hair matrix and the papilla. There is no basement membrane at the dermal–epidermal boundary in the vicinity of the dotted line. *Asterisk* represents a group of intruding dermal cells. The area designated by the *arrows* seems to push the bulb cells. Dorsal skin of a rat at the day of birth was used for the experiment. **B** Formation of spherical papilla. Dorsal skin of a rat at the day of birth was used. **C** Conversion to a cylindrical form. Dorsal skin of a day 7 rat was used. **D** Increase of intercellular space or matrix in the papilla. Dorsal skin of a day 14 rat was used for the experiment. *Bars* 2 µm

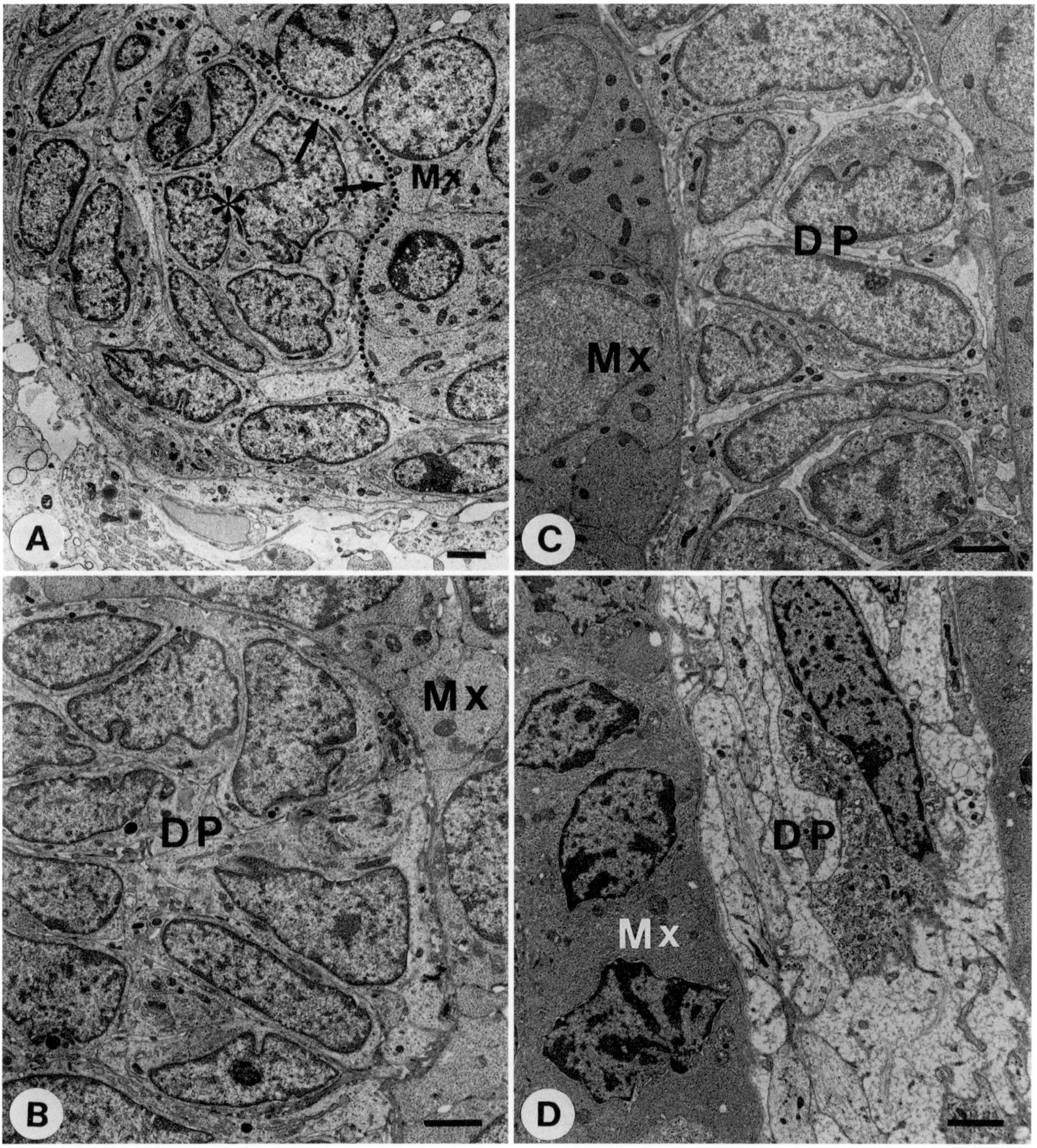

Fig. 6.1.

Before Intrusion

Figure 6.2 shows an immature hair follicle before the intrusion of dermal cells. It consists of inner and outer cells, which are distinguishable by the difference in the orientation of the cells and their nuclei. The boundary of these two cellular groups is drawn as the dotted line in Fig. 6.2. Some inner cells denoted by small asterisks already contain tonofilaments, although these may represent a vestigial event. In ancient mammals this inner cell mass may simply have grown and keratinized to form the hair. The structure, which is composed of four cellular layers as shown in Fig. 6.2 or Fig. 6.3, seems ready for the intrusion of dermal cells to form the papilla. It is interesting to compare the morphology of the connective tissue follicle cells at the side of the hair follicle with those at the tip of the follicle, where the intrusion starts (Fig. 6.3): nuclei are unremarkable in the side area cells, but become polymorphic or indented as the cells approach the intrusion area. I tentatively use the term "intrusion" in this book, though "embracement (by the epidermal follicle cells)" may contribute more to the morphogenesis. The real mechanisms are still ambiguous, but I consider that the active movement of the dermal cells is important.

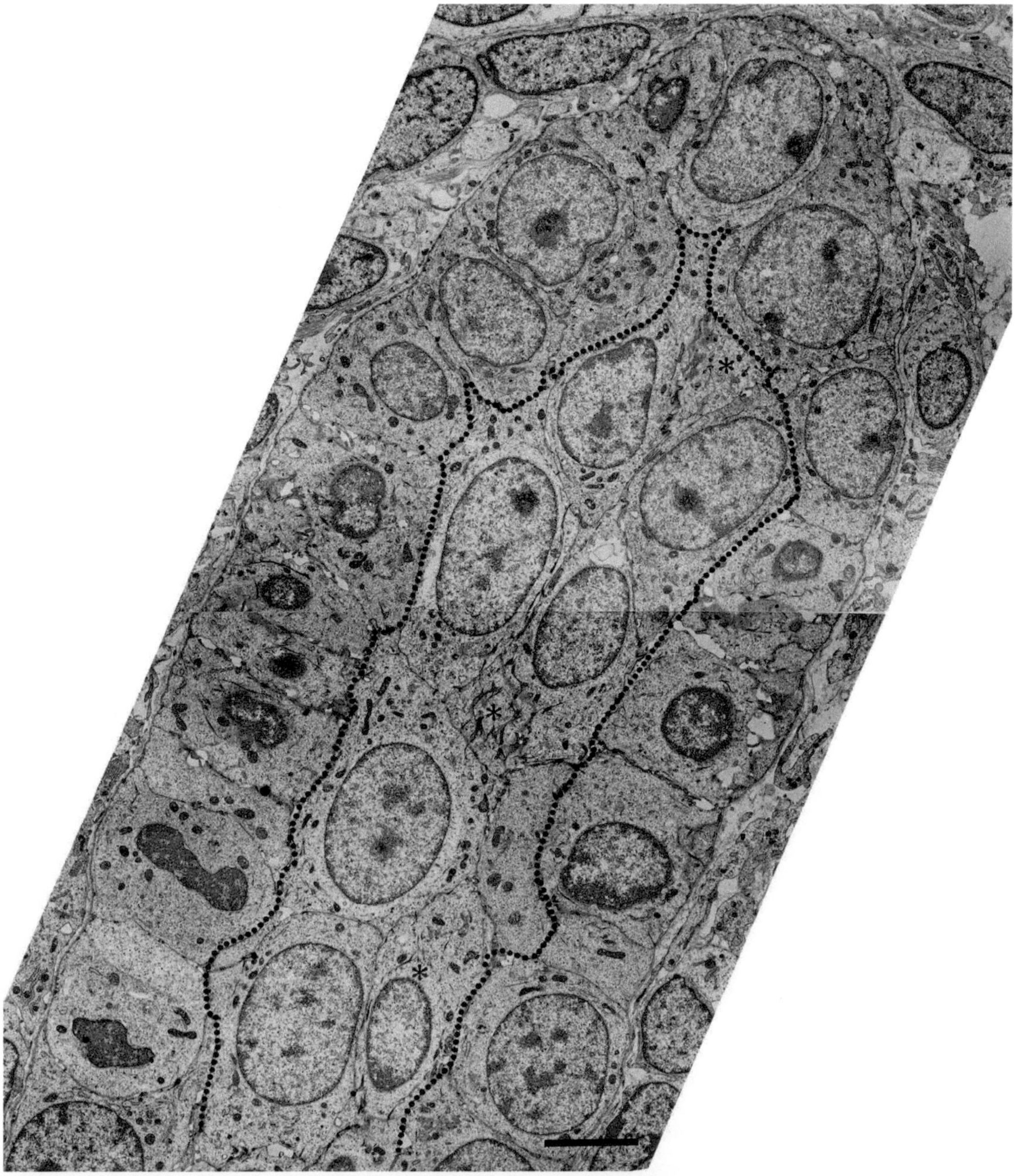

Fig. 6.2. Immature hair follicle before papilla intrusion. Inner cells and outer cells are distinguishable at this stage (stage 2 according to Hardy's diagram; see Fig. 1.3) of the follicle development. The boundary is traced by the *dotted line*. Some cells of the inner area already produce the tonofilament-like structures, as denoted by *small asterisks*, but those may be vestiges. Dorsal skin of a day 7 rat was used for the experiment. *Bar* 5 μm

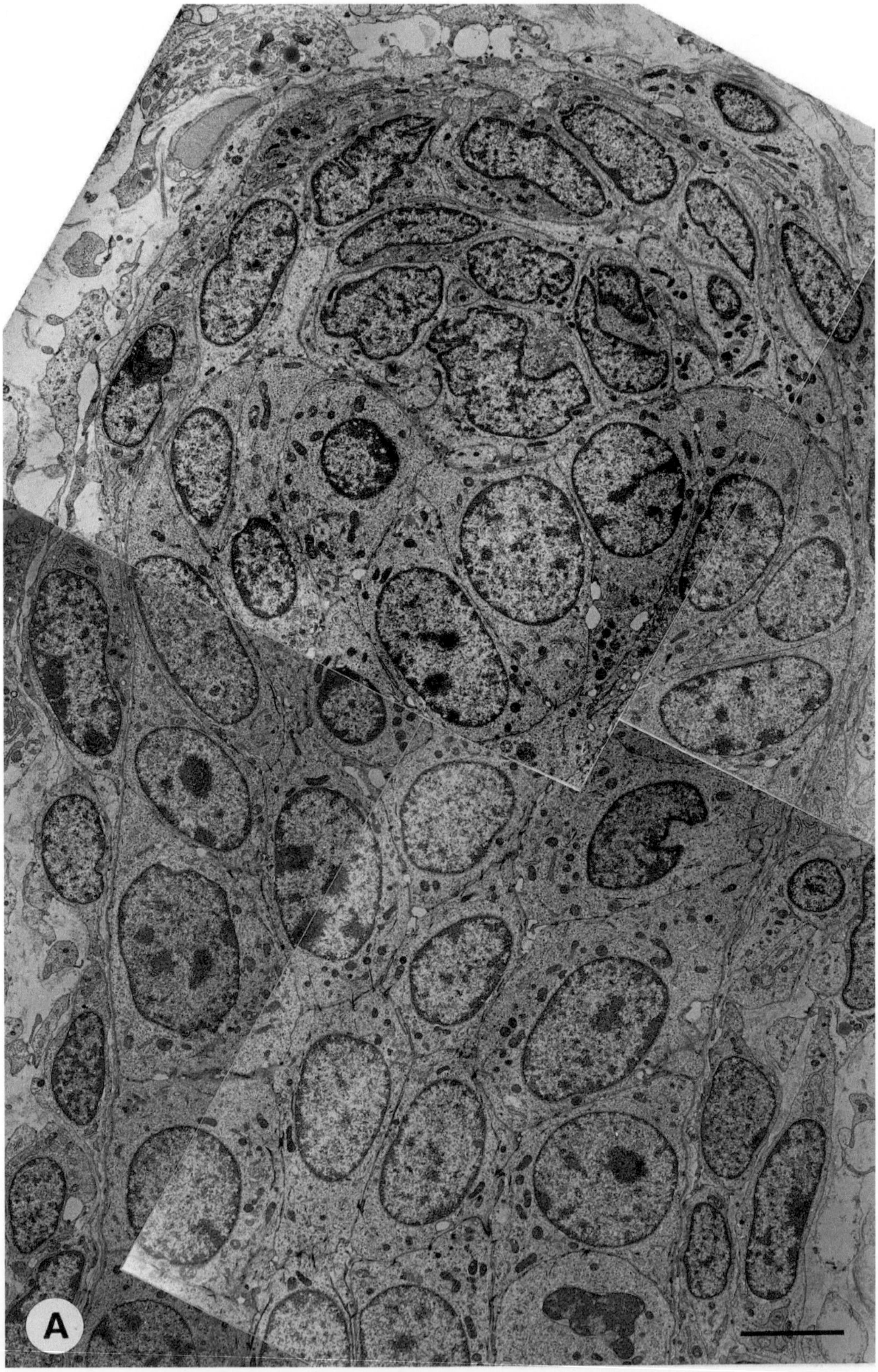

Fig. 6.3.

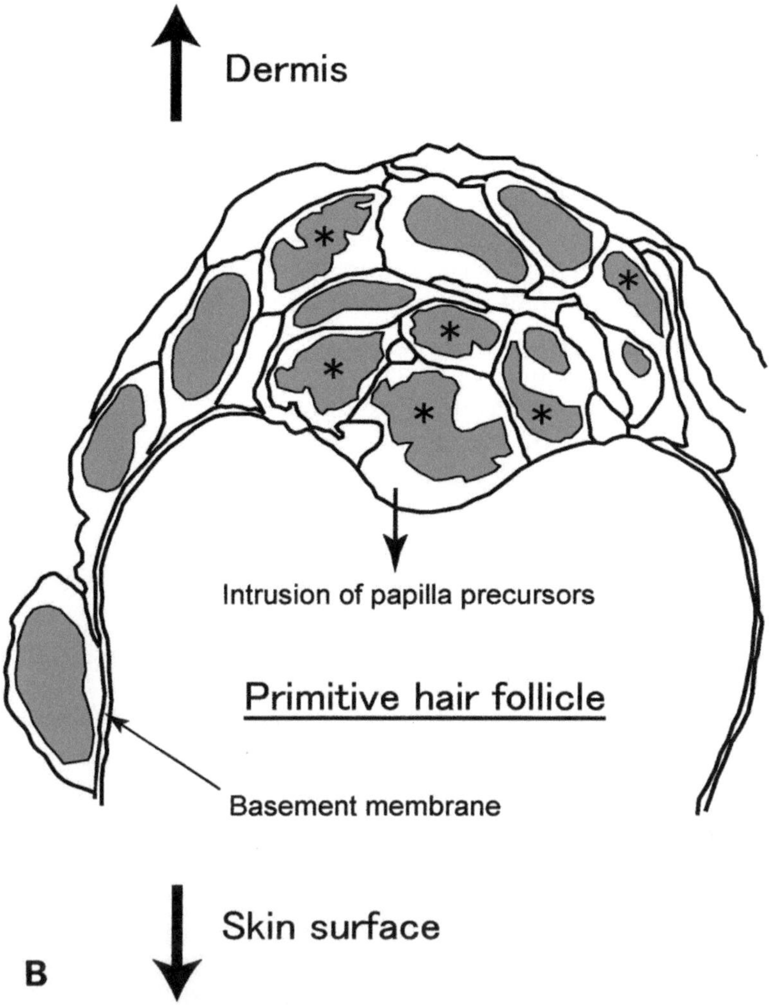

Fig. 6.3. Initial phase of papilla development. **A** An extended view of Fig. 6.1A. The intrusion of the papilla cells is explained in Fig. 6.1. The morphological change of the nuclei of connective tissue follicle cells occurs at the bottom of the hair follicle (*top area* in this figure). Dorsal skin of a rat at the day of birth was used for the experiment. *Bar* 5 μm. **B** Schematic representation of nuclei (*shaded*) of intruding cells. *Asterisks* designate the cells having polymorphic or indented nuclei

First Intrusion Step in the Formation of the Papilla

The opening step in the intrusion of dermal cells into the hair bulb is shown in Figs. 6.1A and 6.3. The cell group marked by the asterisk in Fig. 6.1A as well as the marked cells in Fig. 6.3B seem to be an advanced cohort of papilla precursor cells. The intruding dermal cells are seen to be close to the epidermal bulb cells without the intervention of a basement membrane (see neighborhood of the dotted line in Fig. 6.1A as well as around the intrusion point in Fig. 6.3A). This condition contrasts sharply with ordinary connective tissue follicle cells, which maintain a steady distance from the outer root sheath (e.g., Figs. 4.6, 6.1, 6.2C). In general terms it is natural that dermal and epidermal cells would repulse each other, because selective adhesion (repulsion) is the general basis of tissue identity [8]. The adhesion of dermal and epidermal cells here is therefore a special event. Based on the findings in Figs. 6.1A,B, 6.3, and 6.4A,B, in which cells behave as a group, it is suggested that the intrusion is usually initiated by a mass of inrushing dermal cells rather than any movement of a single independent cell. After the mass invasion of the first group of papilla cells, subsequent cells may enter one by one (Fig. 6.4A).

Figure 6.4 shows the stage just after the intrusion of the first group of cells. The papilla is engulfed by the double-layered epidermal bulb cells. The boundary of the two epidermal layers is shown by a dotted line in Fig. 6.4A. The inner layer probably gives rise to the hair matrix while the outer layer becomes the outer root sheath. I believe that the outer and inner epidermal layers, four cellular columns in longitudinal section, are needed for the intrusion. It is interesting that, after intrusion, the papilla cells again maintain a distance from the matrix cells and seem to communicate with them by means of pseudopodia (Figs. 6.5, 6.6). This concept is physiologically reasonable given that the papilla cells must advance inward to allow sufficient space for succeeding cells wanting to enter the cavity. If affixed junctions were formed between the papilla and matrix, the advance of the papilla cells would be restricted. In contrast to the first cohort, the succeeding cells do not adhere to the epidermal bulb cell even at the position of the entrance (Fig. 6.4A; see upper end).

Fig. 6.4. The foremost inrushing group of dermal papilla cells. *DP*, dermal papilla; *Mx*, hair matrix. The *dotted line* in **A** denotes the boundary between outer and inner layer of the hair bulb cells. An enlarged dermal papilla is shown in **B**. Samples were obtained from the dorsal skin of a rat on day 7 after birth. *Bars* **A** 2 μm; **B** 1 μm

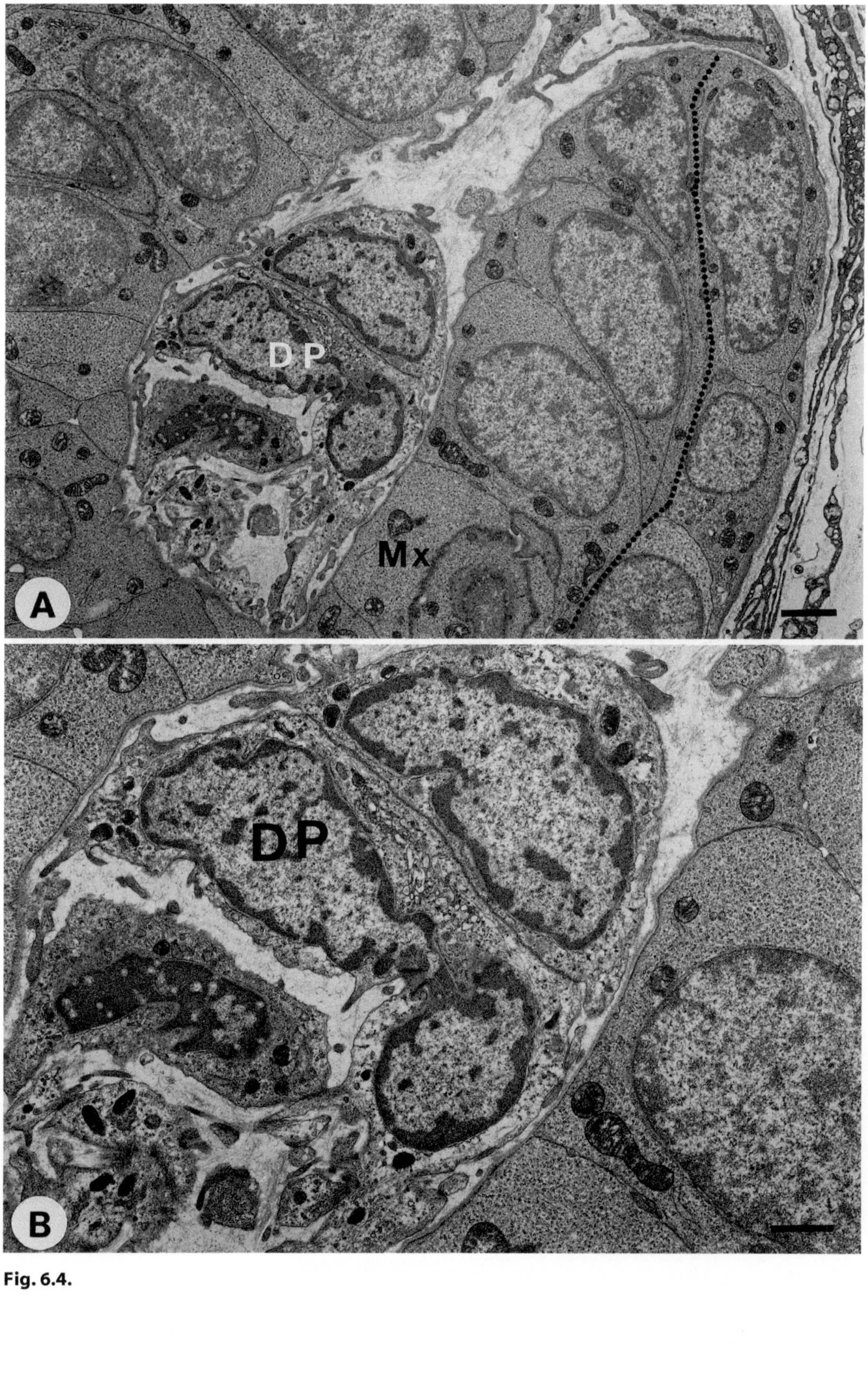

Fig. 6.4.

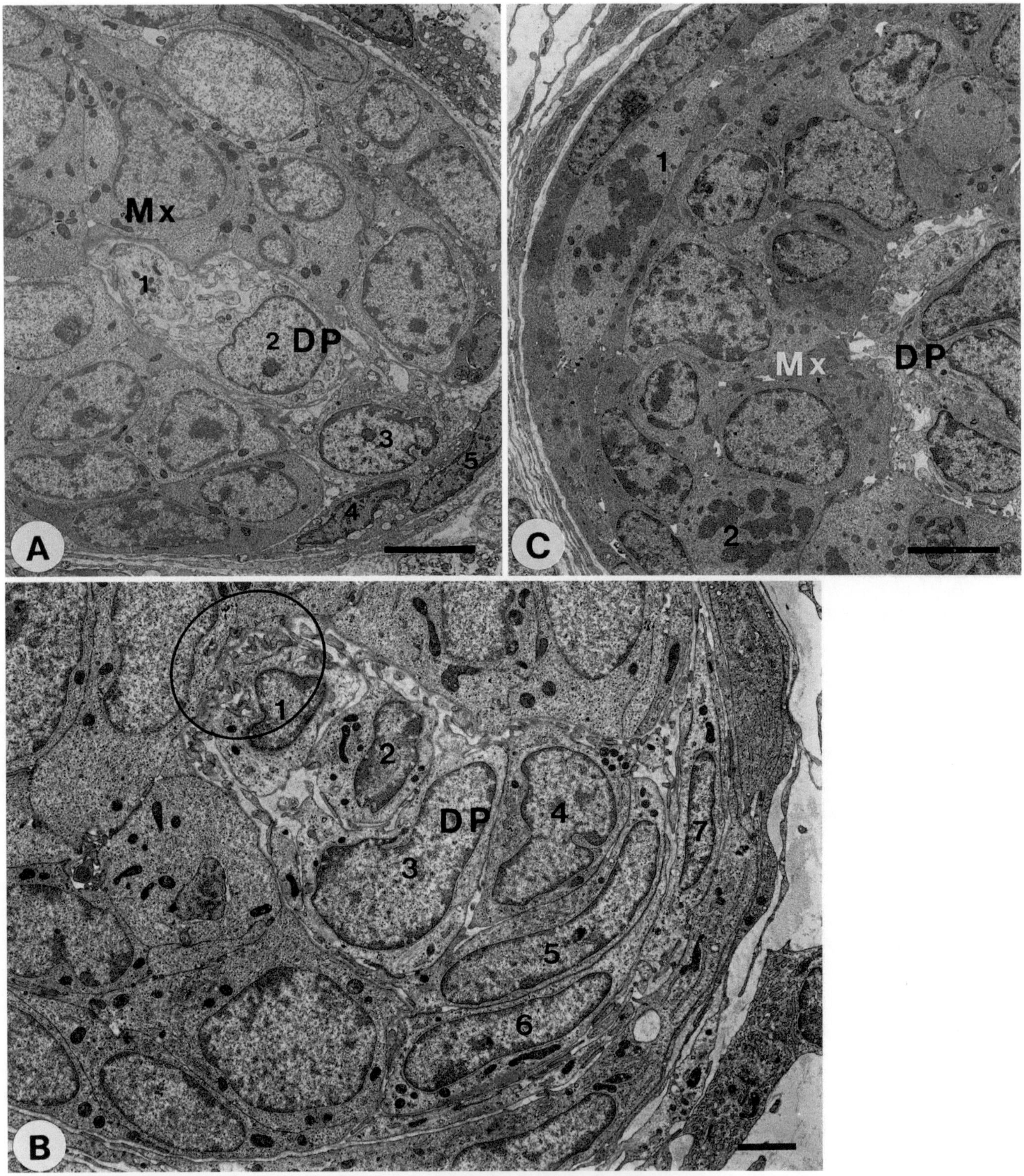

Fig. 6.5. Difference of morphology between preintrusion and postintrusion of the dermal papilla cells and the cell division of bulb cells surrounding the dermal papilla. *DP*, dermal papilla; *Mx*, hair matrix. Numbered cells *1–5* in **A** and *1–6* in **B** show the order of intrusion; *1, 2* in **C** are dividing cells. Samples were obtained from the dorsal skin of day 7 rats (**A**, **B**) and the facial skin of day 2 rats (**C**). *Bars* **A**, **C** 5 μm; **B** 2 μm

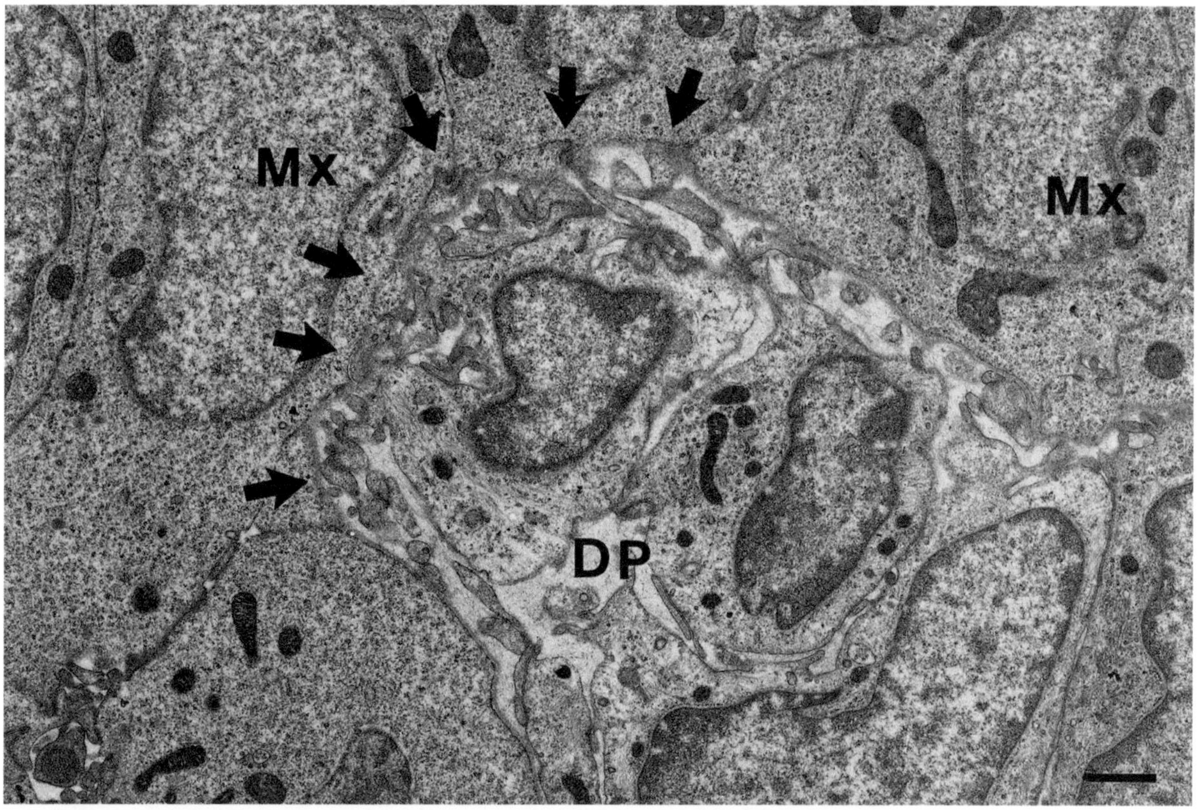

Fig. 6.6. Pseudopodia of dermal papilla cells. *DP*, dermal papilla; *Mx*, hair matrix; *arrows*, points of contact between the surface of the hair matrix cells and pseudopodium of the papilla cells. Dorsal skin of a day 7 rat was used for the experiment. *Bar* 1 μm

Characteristics of the Dermal Papilla Cells

The intruding or early papilla cells are characterized by the polymorphic or indented structure of their nuclei (Figs. 6.1, 6.3–6.5). At the initial stage of papilla formation, the two layers surrounding the bulb cells can be distinguished into outer and inner layers according to the orientation of the cell and its nucleus (Fig. 6.2). As the stage proceeds, the morphological difference between these two layers becomes more evident (Fig. 6.5C). The hair bulb in Fig. 6.5C consists of two outer layers of flat cells, two inner layers of cuboidal cells, and the dermal papilla at the center. This image includes views of cell division in both outer and inner layer cells. The former is probably a division of the cell belonging to the companion layer (designated as 1 in Fig. 6.5C).

The numbers given to the cells in Fig. 6.5A,B denote the order of entrance into the papilla cavity, showing that the flat cells with polymorphic nuclei become cuboidal as they advance into the papilla cavity. The front surface of the leading cell (numbered 1 in Fig. 6.5B) seems to be actively generating pseudopodia. The area encircled by the ellipse in Fig. 6.5B is enlarged in Fig. 6.6. The arrows in Fig. 6.6 indicate the area where pseudopodia formation is active. Figure 6.6 suggests that communication between dermal papilla cells and epidermal hair matrix cells is executed through a system mediated by pseudopodia, which may affect the growth and differentiation of hair matrix cells.

The dermal papilla cells are allowed to approach the matrix cells very closely at the surface of the pseudopodia, forming what appears to be a kind of tight junction between dermal and epidermal cells (white arrows in Fig. 6.7). The presence of physical, nutritional, and informational streams from papilla to matrix is likely (thick arrow in Fig. 6.7C shows the direction of the stream). Even after entering the papilla cavity, some dermal cells retain a polymorphous nucleus. As stated earlier, the shape of the papilla changes to cylindrical. The papilla in the dorsal skin of a rat at day 14 after birth consists largely of single lined but separate cells and abundant intercellular matrix (Fig. 6.8A,B). Some papilla cells contain rough endoplasmic reticula (Fig. 6.8C), which are rarely observed in hair matrix cells.

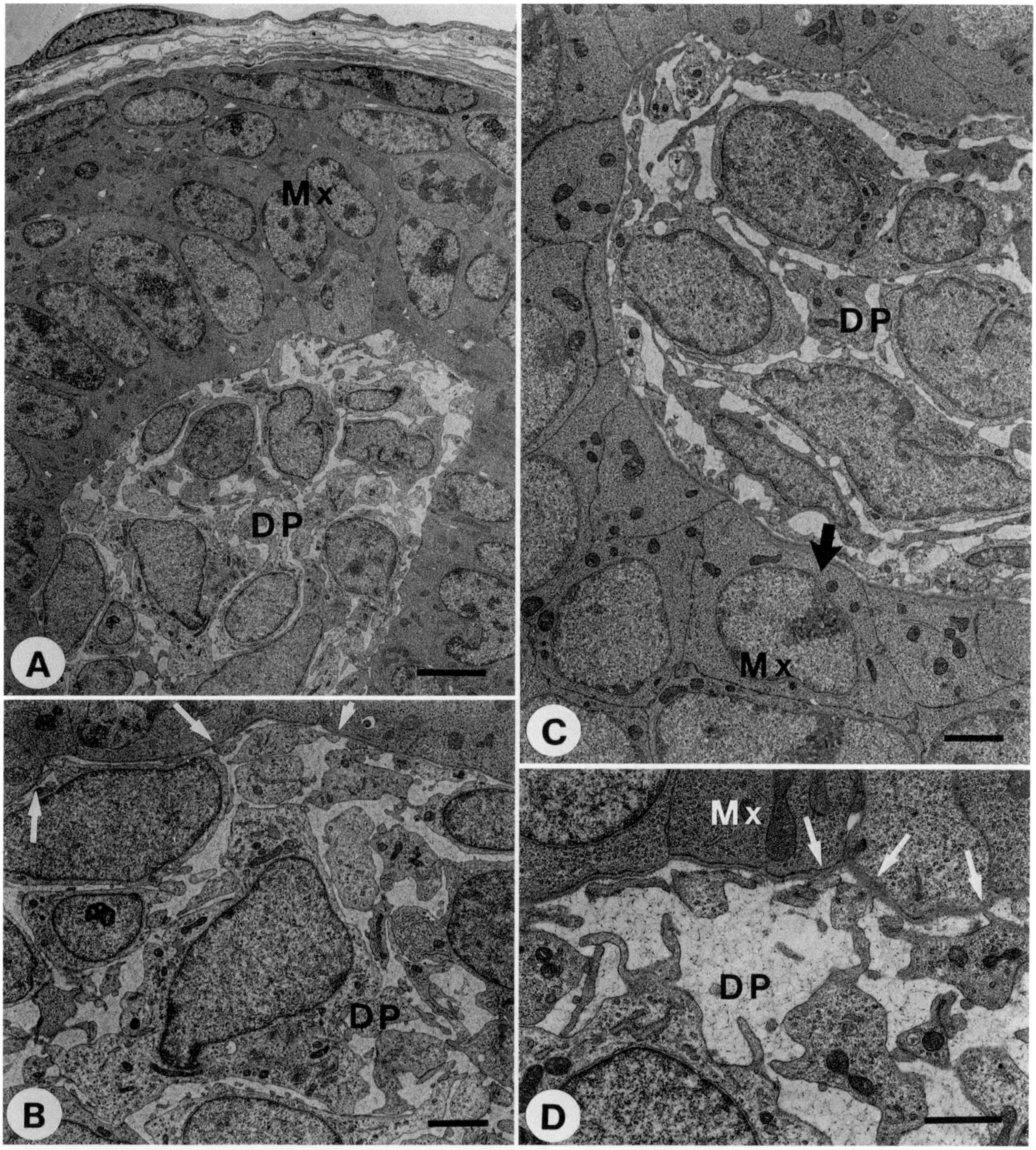

Fig. 6.7. Sagittal section of hair bulb to show the morphology of hair bulb and dermal papilla cells. *DP*, dermal papilla, *Mx*, hair matrix. In the areas pointed to by the *white arrows* in **B** and **D**, the papilla cells seem to contact the matrix cells without the intervention of the extracellular matrix. The *thick arrow* in **C** represents presumptive informational and nutritional flow. Samples were obtained from the dorsal skin of day 2 (**A**, **B**) and day 7 (**C**, **D**) rats. *Bars* **A** 5 μm; **B–D** 2 μm

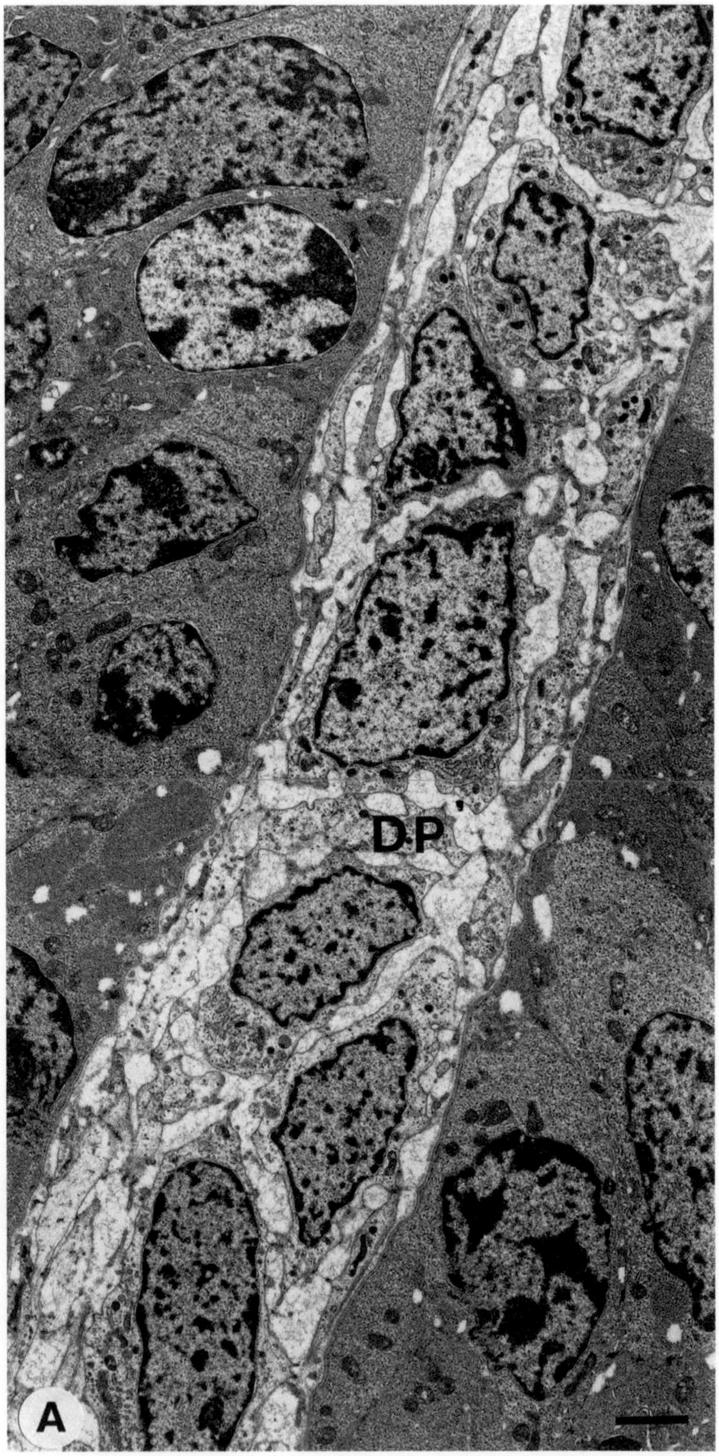

Fig. 6.8.

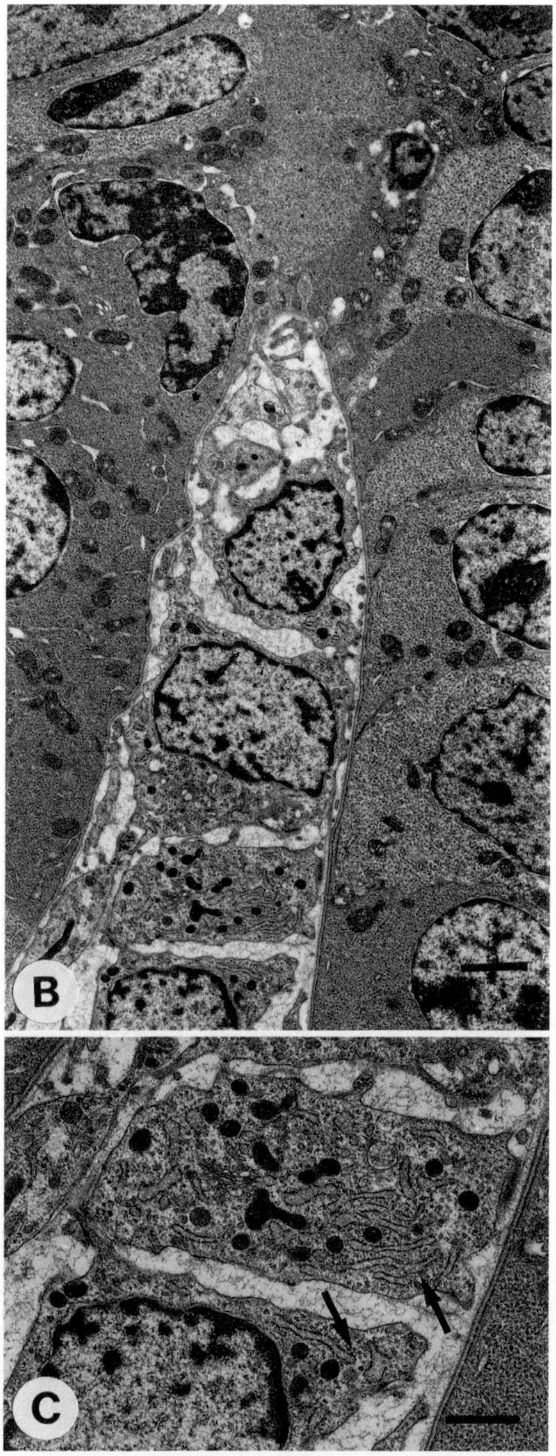

Fig. 6.8. Cylindrical papilla. *DP*, dermal papilla. The upper end of **A** continues to the lower end of **B**; **C** is an enlarged view of the lower part of **B**. *Arrows* in **C** denote the rough endoplasmic reticulum. Dorsal skin of day 14 rats was used for the experiment. *Bars* **A, B** 2 μm; **C** 1 μm

Hair Matrix

Figure 6.9 shows the cross section of a hair bulb in which Henle's layer exhibits no signs of differentiation. The bulb consists of four or five layers. The orientation of the cells of the inner two and outer two layers is clearly distinguishable, but that of the third layer is obscure. The outermost layer (layer five) is the outer root sheath whereas layer four is thought to be the companion layer. The two innermost layers are thought to be the hair matrix. The third layer is probably included in the root area of the inner root sheath. The epidermal bulb is encircled by several layers of dermal cells, referred to as the connective tissue follicle (CTF). The cross section of an upper level is shown in Fig. 6.10. At this level, Henle's layer starts to differentiate. Some trichohyalin granules are observed in the outermost layer cells of the inner root sheath (IRS) area. The two other layers of the IRS shown here are assumed to be Huxley's layer and the cuticle of the inner root sheath. The dermal papilla does not appear at this height. The outermost cells in the hair shaft (HS) area exhibit distinct orientation compared with the inner cells, suggesting that these cells are the precursors of hair cuticle cells.

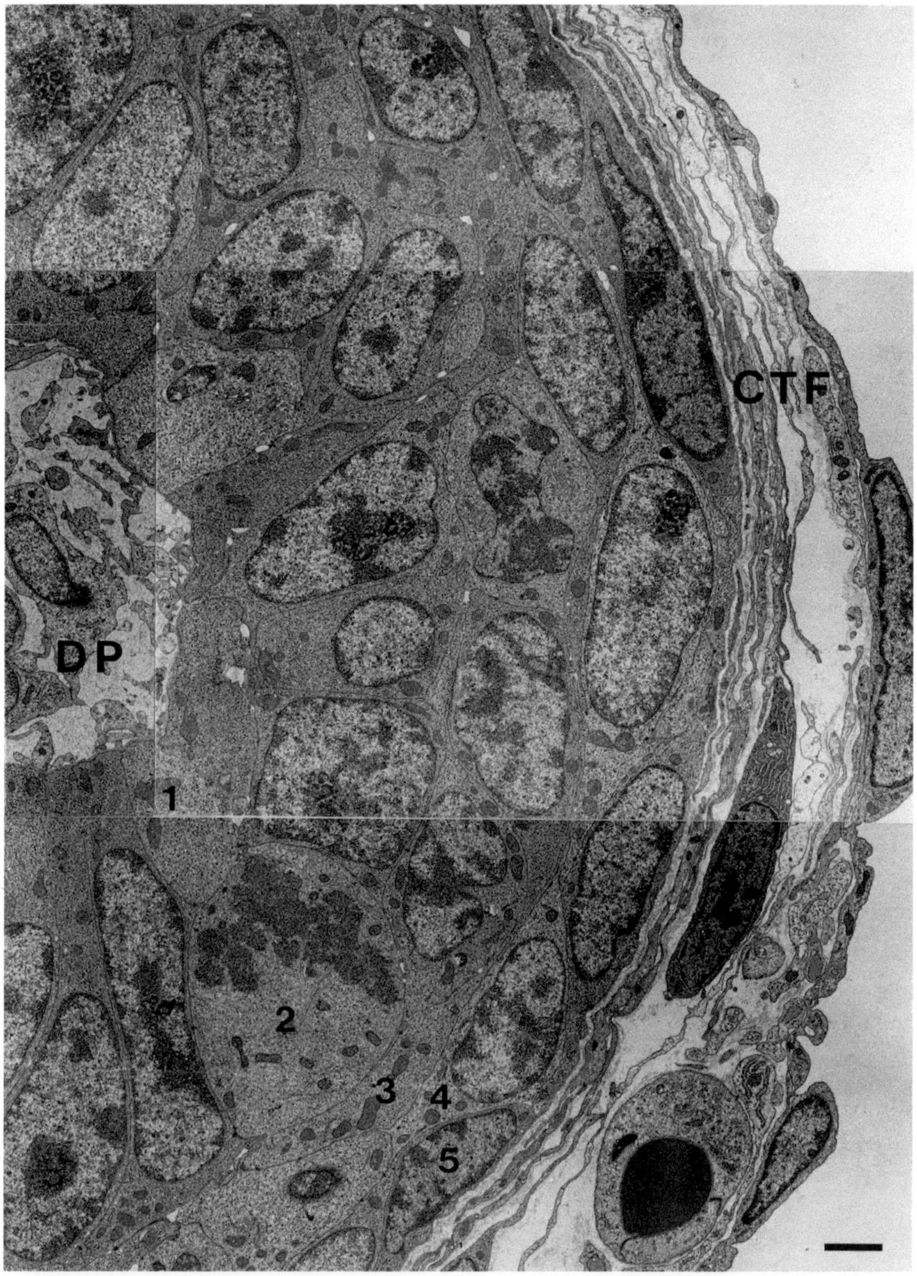

Fig. 6.9. Cross section of hair bulb at the level of upper part of dermal papilla. *DP*, dermal papilla; *CTF*, connective tissue follicle. This section exhibits five concentric hair bulb cellular layers, numbered *1* to *5* from inside to outside. Dorsal skin of day 2 rats was used for the experiment. Bar 2 μm

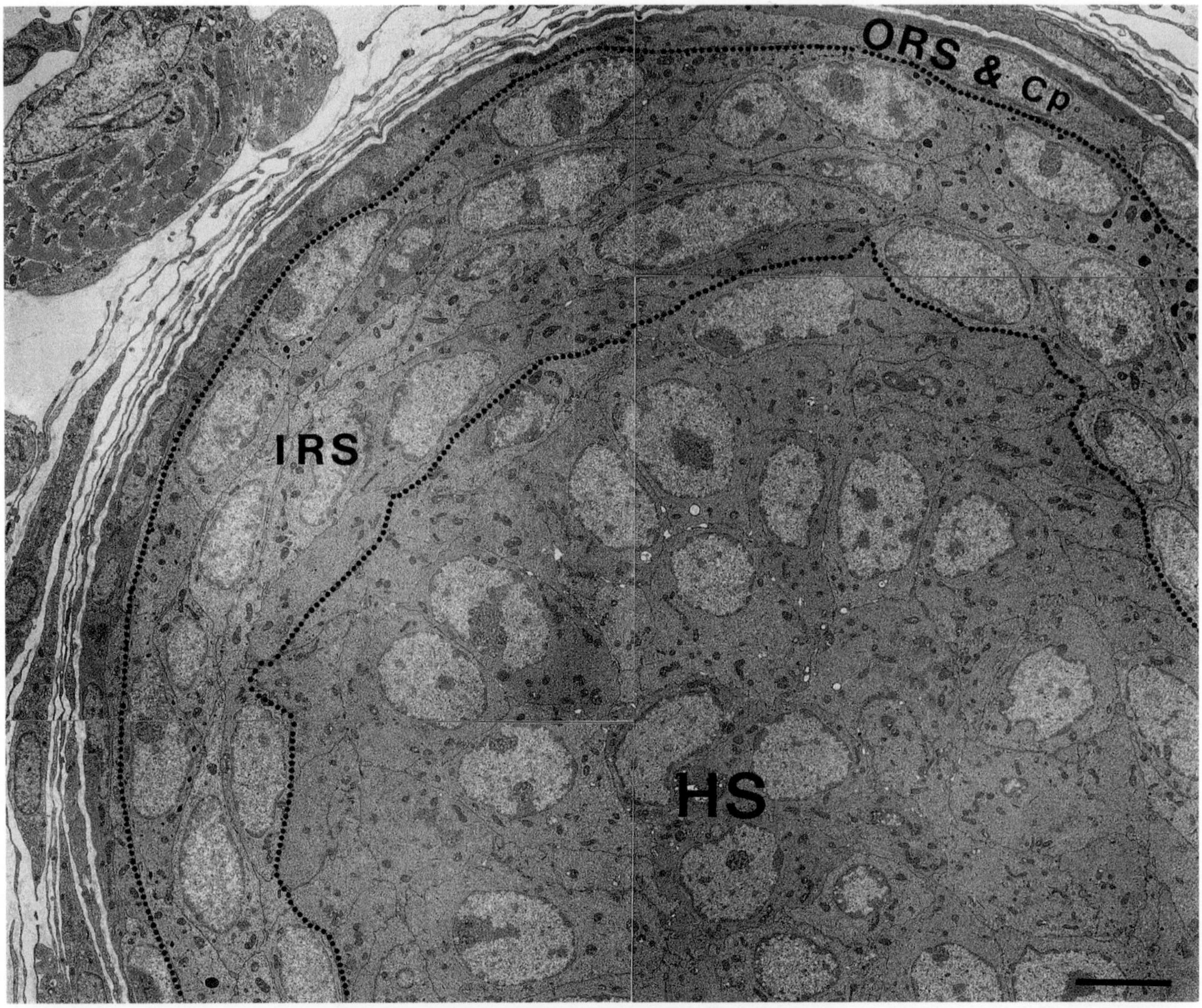

Fig. 6.10. Cross section of a hair bulb at a level higher than the top of dermal papilla. Precursors of hair shaft and inner and outer sheath cells are distinguishable based on their morphology at this level and are designated as *HS*, *IRS*, and *ORS & Cp*, respectively, in this figure. Facial skin of day 2 rats was used for the experiments. *Bar* 5 μm

Recruitment and Evacuation of Stem Cells

Human scalp hairs grow an average of about 10 mm/month. Considering the size of hair cells (see Fig. 1.6), which range from about 2 to 20 μm in height, this is rapid growth indeed. According to simple calculation, the length of 10 mm corresponds to 500 to 5,000 cells, meaning that 17 to 170 cells must be produced per day in one vertical cellular lane. The sensory hair of the neonatal rat as well as the human beard may grow 1–1.5 mm/day, equivalent to 50 to 750 cells/lane/day. It is therefore impossible for one lane of each tissue in hair to be produced from one basal (root) stem cell, as occurs in the tapeworm, for example. I present one possible answer to this quandary in Fig. 6.11A. In this model, (1) matrix cells enter the cell cycle in the primary niche or "matrix" in Fig. 6.11A. (2) At the late G_2 phase, cells are recruited to the basal position of the lanes, assumed to be the secondary niche or "root" in Fig. 6.11A. (3) The stem cell arrives at its secondary niche, promptly enters the M phase, and executes an unequal cell division. (4) The upper daughter cell produced by the unequal cell division undergoes differentiation in accordance with its position (i.e., Huxley's layer type, Henle's layer type, and so on), while the lower daughter cell does not enter into the differentiation pathway, but rather immediately evacuates the niche for the next stem cell in late G_2 phase, and returns to the primary niche.

A group of stem cells may constitute cellular chains and move as a "wheel" (Fig. 6.11B). In this case, the movement of the wheel would be synchronized with the cell cycle. Alternatively, individual stem cells may stay, enter the cell cycle at the primary niche, and migrate (go and return) only around the time of the M phase to divide at the secondary niche, in the manner of a shuttle (Fig. 6.11B). The dermal papilla probably instructs the matrix cells to execute the cell cycle. Differentiation may be controlled by the outer root sheath, other neighborhood cells, or dermal papilla cells. Of course, the problem of regulation is not fully understood, and the explanation just given includes several hypothetical processes.

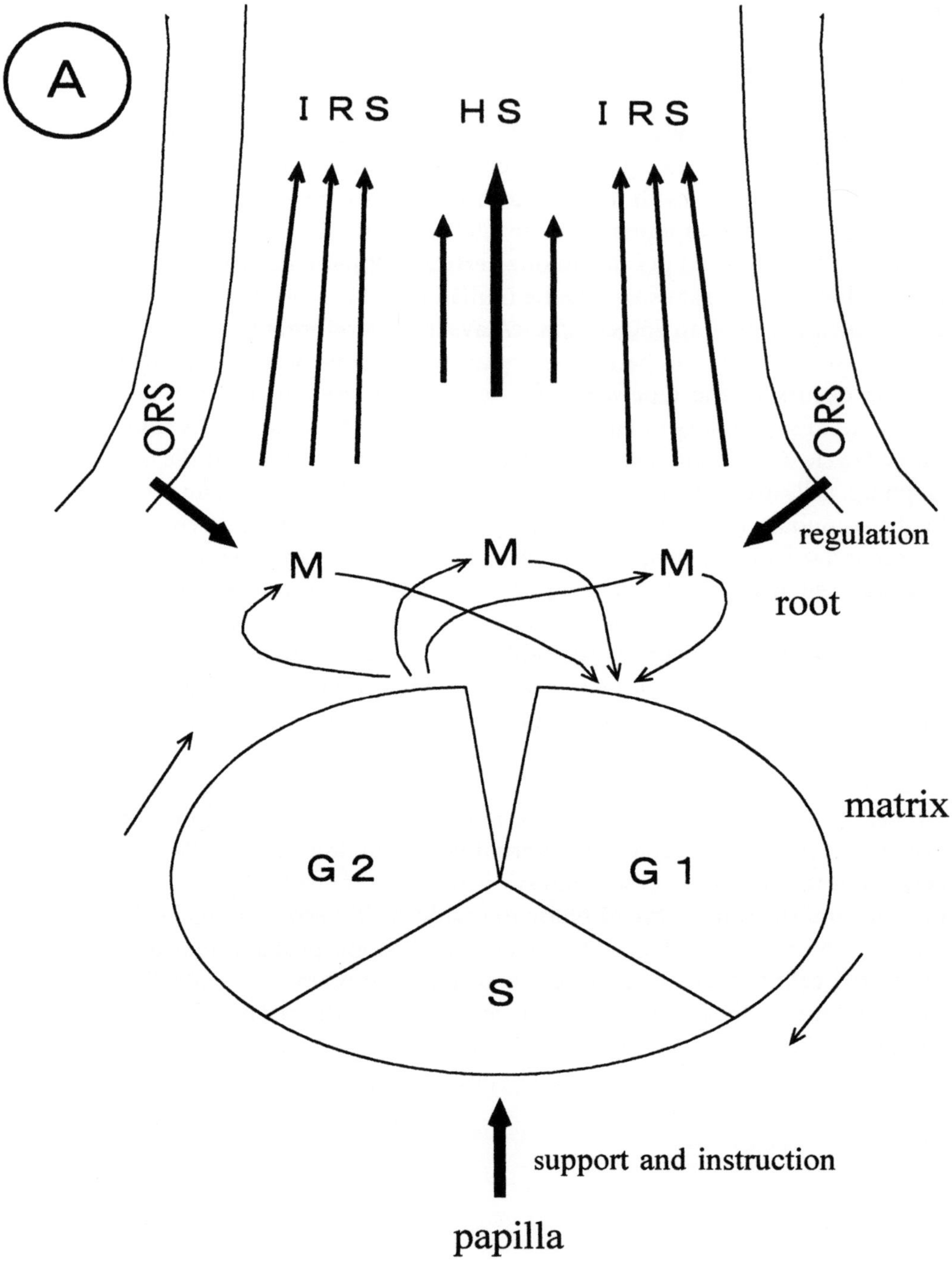
A
IRS
HS
IRS
ORS
ORS
regulation
M
M
M
root
matrix
G 2
G 1
S
support and instruction
papilla

Fig. 6.11.

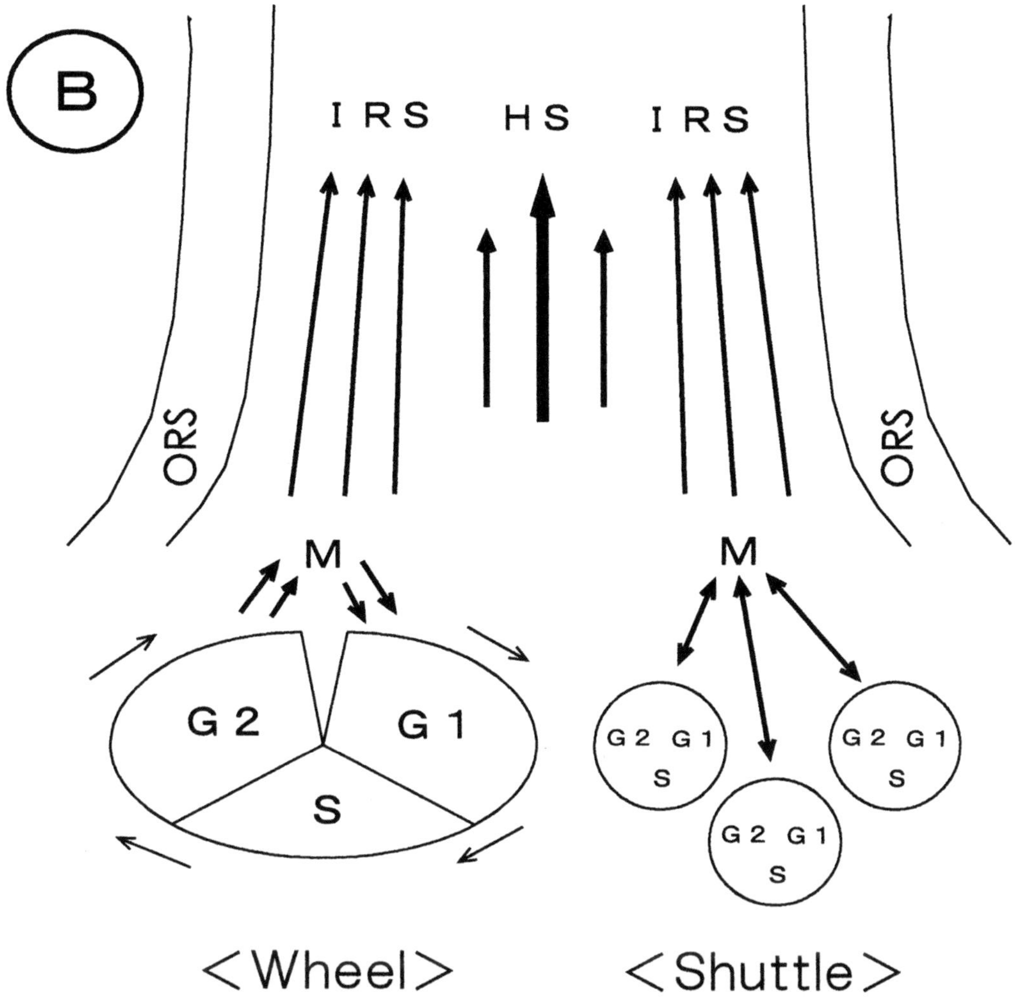

Fig. 6.11. An explanation of the mechanisms that produce hair follicle cells. *G1*, *S*, *G2*, and *M* represent the cell cycle phases. One fixed stem cell at the root of the cellular column is not enough to maintain the hair growth, as stated in the text. **A** Some late G_2-phase cells in the matrix are recruited to the basal (root) areas of *IRS* (inner root sheath) and *HS* (hair shaft), and at these secondary niches, the cells enter into M phase and execute unequal cell divisions that give rise to both differentiation-oriented cells and reproduced stem cells. The former cells create IRS and HS tissues, and the latter early G_1-phase cells return to the primary niche in the matrix to undergo the cell cycle again. These processes may be controlled by *ORS* (outer root sheath) and the dermal papilla as well as matrix cells. **B** There are two possible pathways for stem cells to support high-speed growth of the hair tissues. In the "wheel" hypothesis, stem cells are imagined to run in the cell cycle circuit, and the late G_2-phase cells (for example) are trapped at the second niche (bottom of the cellular column). The circuit is the primary niche of the stem cells in this case. By contrast, stem cells in the "shuttle" hypothesis do not run in the circuit, but stay at the primary niche, and when they enter the G_2 phase they are recruited to the second niche. After unequal cell division at the secondary site, the daughter stem cell returns to the primary niche

References

1. Durward A, Rudall KM (1958) The vascularity and patterns of growth of hair follicle. In: Montagna W, Ellis RA (eds) The biology of hair growth. Academic Press, New York, pp 189–218
2. Sholley MM, Cotran RS (1976) Endothelial DNA synthesis in the microvasculature of rat skin during the hair growth cycle. Am J Anat 147:243–254
3. Mecklenburg L, Tobin DJ, Muller-Rover S, Handjiski B, Wendt G, Peters EMJ, Moll I, Paus R (2000) Active hair growth (anagen) is associated with angiogenesis. J Invest Dermatol 114:909–916
4. Roth SI, Helwig EB (1964) The cytology of the dermal papilla, the bulb, and the root sheaths of the mouse hair. J Ultrastruct Res 11:33–51
5. Morioka K, Sato-Kusubata K, Kawashima S, Ueno T, Kominami E, Sakuraba H, Ihara S (2001) Localization of cathepsins B, D, L, LAMP-1 and μ-calpain in developing hair follicles. Acta Histochem Cytochem 34:337–347
6. Hardy MH (1992) The secret life of the hair follicle. Trends Genet 8:55–61
7. Randall VA, Hibberts NA, Thornton MJ, Hamada K, Merrick AE, Kato S, Jenner TJ, De Oliveira I, Messenger AG (2000) The hair follicle: a paradoxical androgen target organ. Hormone Res 54:243–250
8. Gilbert SF (2003) Developmental biology, 7th edn. Sinauer Associates, Sunderland, MA, pp 69–77

Appendix I

Cells and Tissues That Surround the Hair Follicle

Fibroblasts and Mast Cells

The fibroblast, designated as 2 in Fig. AI.1, is a major cellular member of the dermis. Distinctive characteristics of dermal fibroblasts are their extended form with an amorphous cellular contour, abundant swollen rough endoplasmic reticulum (ER) (b in Fig. AI.1) in the cytoplasm, and an indented nucleus. Their major role is to secrete the components of the extracellular matrix such as collagens, fibronectins, proteoglycans, and other dermal matrix components. However, some fibroblasts or related cells may specialize to induce hair follicle formation, engulf the hair bulb to make the dermal sheath, or intrude into the hair bulb to form the papilla [1, 2]. The structure of the connective tissue follicle in the mystacious (sensory) hairs is particularly complex, since they are heavily vascularized and innervated. Dermal fibroblasts are therefore considered to play an important supportive role in hair formation.

The cell located at the right side of the fibroblast in Fig. AI.1 is a mast cell, designated 3. This cell is distinctive with regard to its inclusion of electron-dense histamine granules (arrow c in Fig. AI.1). Mast cells and basophils are activated by immunological stimuli, such as immunoglobulin E (IgE) challenge [3, 4]. The contents of their granules, mainly histamine, are released at once in a process known as compound exocytosis, which then evokes the inflammatory reaction. During this event granules fuse with one another, forming channels to the cell surface [5] (Fig. AI.2). These channels (arrows in Fig. AI.2) and special membrane folds (arrowheads in Fig. AI.2) are characteristic of activated mast cells.

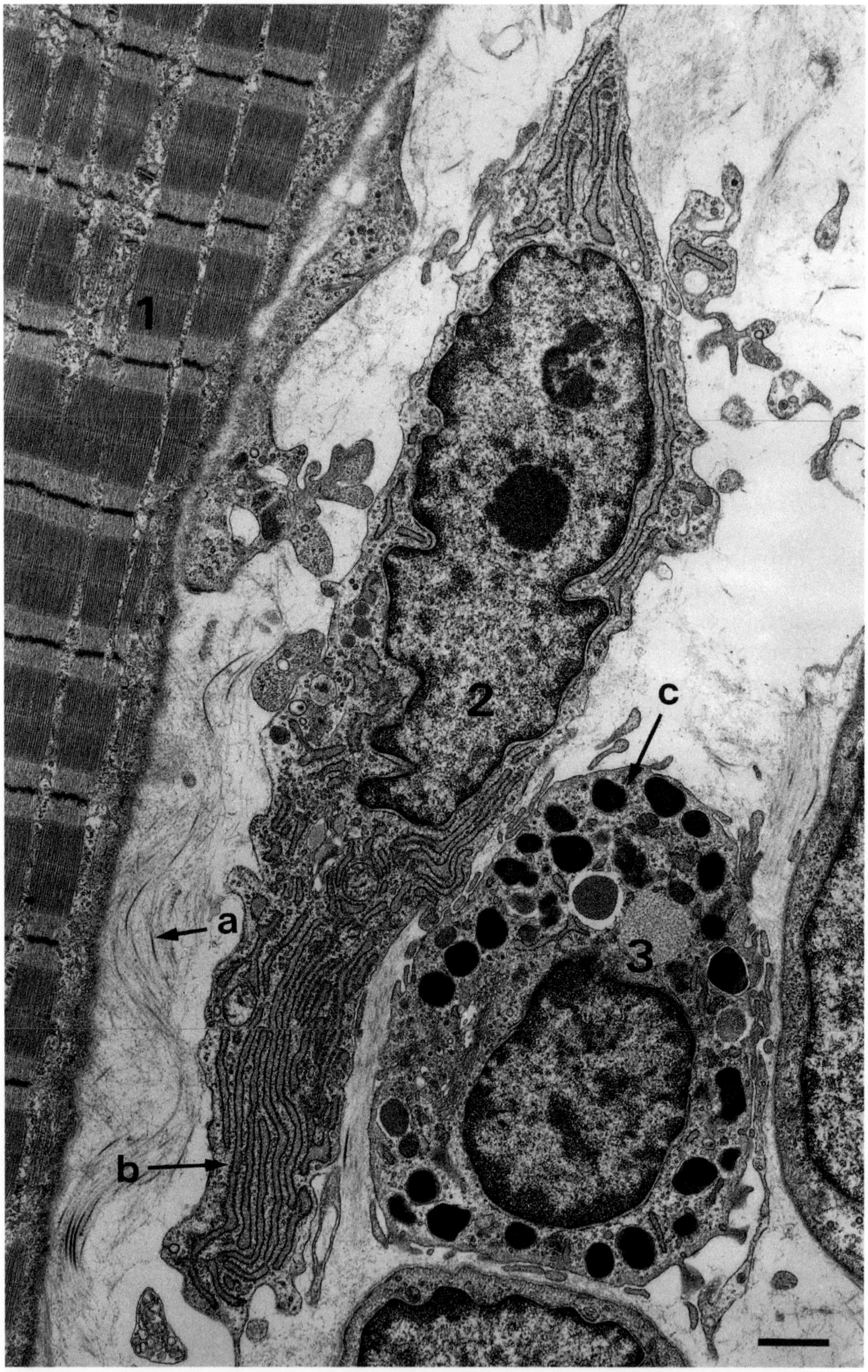

Fig. AI.1. Fibroblast, mast cell, and dermal muscle cell. *1*, Dermal muscle cell; *2*, fibroblast; *3*, mast cell; *a*, collagen fibers; *b*, rough endoplasmic reticulum; *c*, electron-dense granule that contains histamine. The fibroblast has an indented nucleus and exuberant rough endoplasmic reticulum. The mast cell that has not been activated contains many electron-dense granules, which include histamine. Dermal muscle cells are also shown in Figs. AI.3–AI.6. Dorsal skin of a day 14 rat was used for the experiment. *Bar* 1 μm

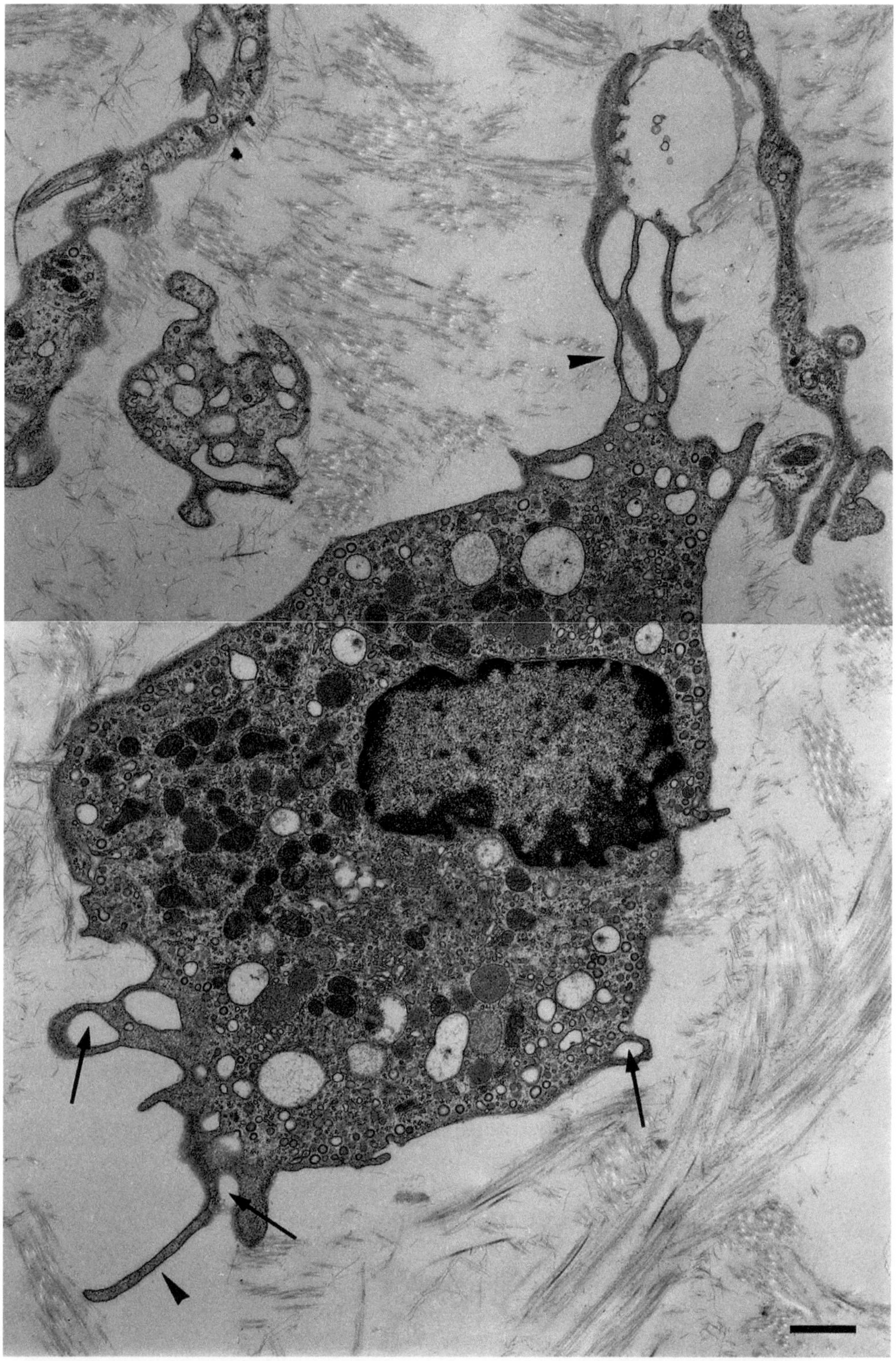

Fig. AI.2. Activated mast cell. The contents of the granules found in Fig. AI.1 were released by exocytosis. *Arrows*, space or canal formed by the compound exocytosis; *arrowheads*, membrane folds. Dorsal skin of a day 14 rat was used for the experiment. *Bar* 1 μm

Dermal Muscle

Another cell appearing in Fig. AI.1 is the muscle cell designated as 1. Most mammals have dermal muscle under the skin to disperse insects or remove water droplets [6, 7]. This function has been lost during evolution from most parts of the human body. As an exception, the facial muscles have developed for expression, which may reflect intellectual and emotional development. A dermal muscle cell of the dorsal skin of the neonatal rat is shown in Fig. AI.3, showing that the dermal muscle is a kind of striated muscle having contractile units, called sarcomeres. The production of sarcomeres seems to occur rapidly, and mitochondria are not yet been provided in the spaces between sarcomeres. The formation of the musculature then proceeds as shown in Fig. AI.4A. The nuclear membrane has distinctive dents at this stage (see cell 1 in Fig. AI.4A). Figure AI.4A shows an enlarged view of a sarcomere, including H, A, I, and Z bands. The horizontal lines consist of alternating thin and thick filaments, the former composed essentially of F-actin, tropomyosin, and troponin and the latter of myosin. The structure of striated muscle and its molecular interactions are described in detail by Fawcett [8] Among vertical bands, Z bands rich in α-actinin anchor the thin filaments. One unit of the sarcomere is usually considered as the span from one Z band to the next. H bands anchor the thick filaments [5] (Fig. AI.4B). I bands contain only thin filaments whereas H bands contain only thick filaments. A bands contain both. The mechanism of muscle contraction is described elsewhere [9].

Figures AI.5 and AI.6 show the process of muscle formation. Figure AI.5A–C and the lower right-hand corner of Fig. AI.6A suggest the formation of alternate thin and thick filaments without vertical (H, A, I, Z) bands, although Z band-like structures are sometimes formed in the early period. Formations of Z bands are seen in Fig. AI.6. Tracing from 1 to 3, 4 to 6, and 7 to 9 in Fig. AI.6A as well as from 1 to 4 and 5 to 8 in Fig. AI.6B, it is evident that the H band is formed in the center of the immature sarcomere, which consists solely of horizontal filaments before the formation of the H band. In contrast, the Z bands are formed in the space between the filament bundles. The process from 1 to 4 in Fig. AI.6B suggests that the anchoring of thin filaments to the Z band is carried out just after completion of the Z band. The thin filaments do not appear near the Z band when it (the Z band) is under development. For example, number 3 in Fig. AI.6B shows that an upper half naked (without association of thin filaments) Z band under construction and a lower half thin filament-associated Z band that seems to have been just completed. At the stage shown in Fig. AI.6, mitochondria localize between the shelves of connected sarcomeres. A second dermal muscle type is the smooth muscle termed "arrector pili," which is responsible for causing the hair to stand. This apparatus is probably not extensively developed in the neonatal period.

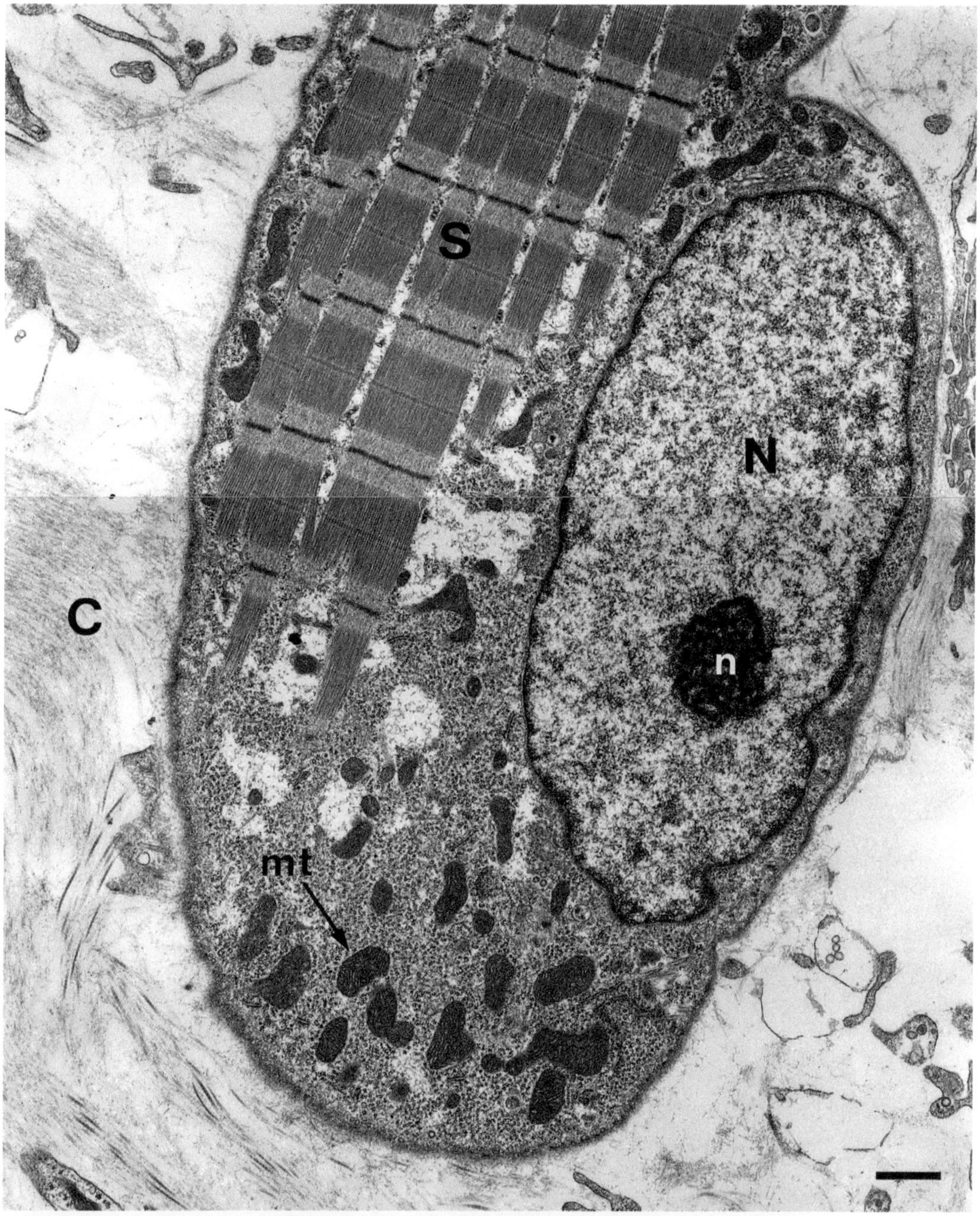

Fig. AI.3. Premature dermal muscle cell. *C*, collagen fibers in the dermal matrix; *s*, sarcomeres; *N*, nucleus of the muscle cell; *n*, nucleolus; *mt*, mitochondrion. This figure suggests that the formation of the muscle structure extended the shape of the cell. The nucleus is not indented. Dorsal skin of a day 14 rat was used for the experiment. *Bar* 1 μm

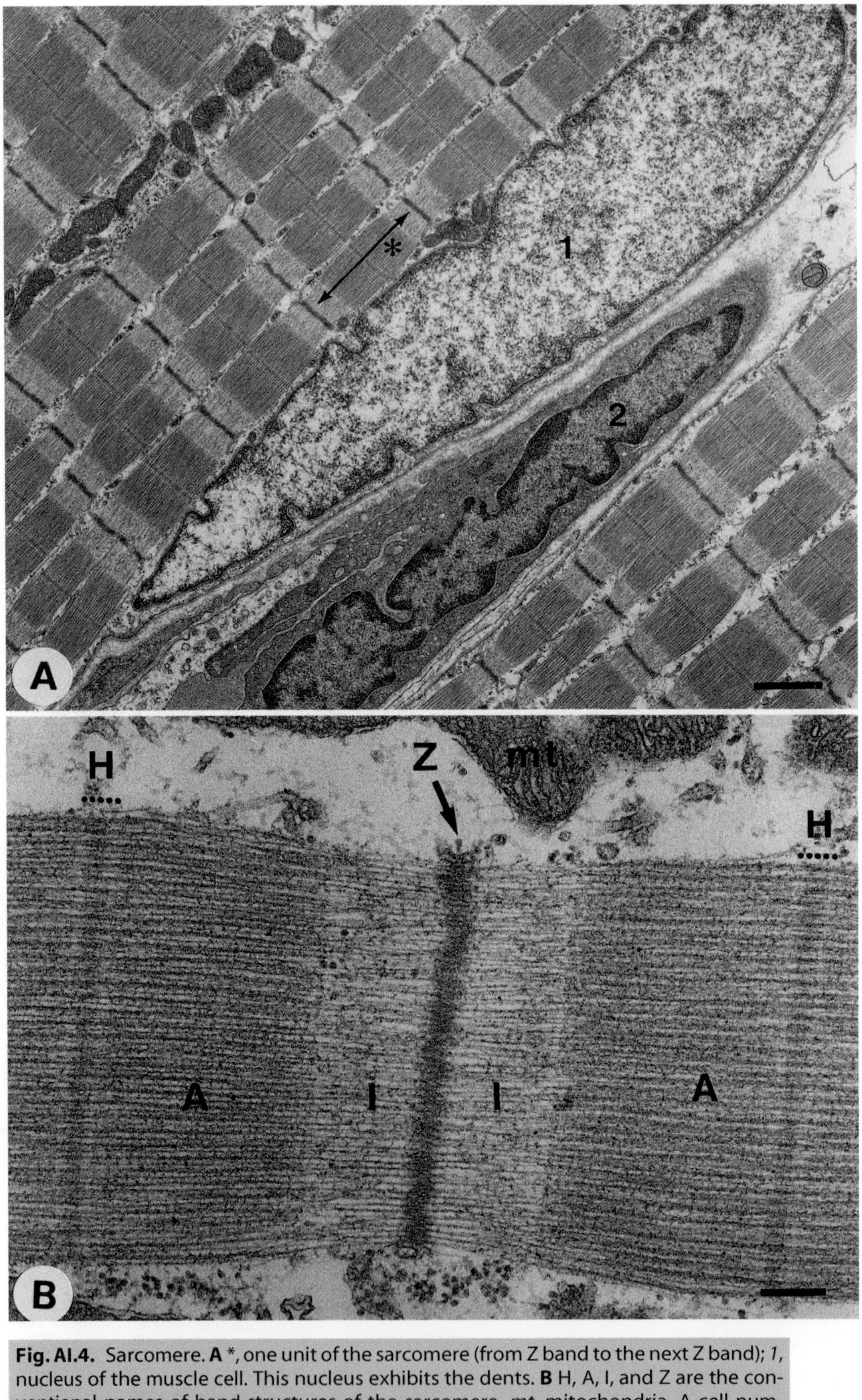

Fig. AI.4. Sarcomere. **A** *, one unit of the sarcomere (from Z band to the next Z band); *1*, nucleus of the muscle cell. This nucleus exhibits the dents. **B** H, A, I, and Z are the conventional names of band structures of the sarcomere. *mt*, mitochondria. A cell numbered *2* is not identified. Dorsal skin of day 14 rats was used for both experiments. *Bars* **A** 1 µm; **B** 200 nm

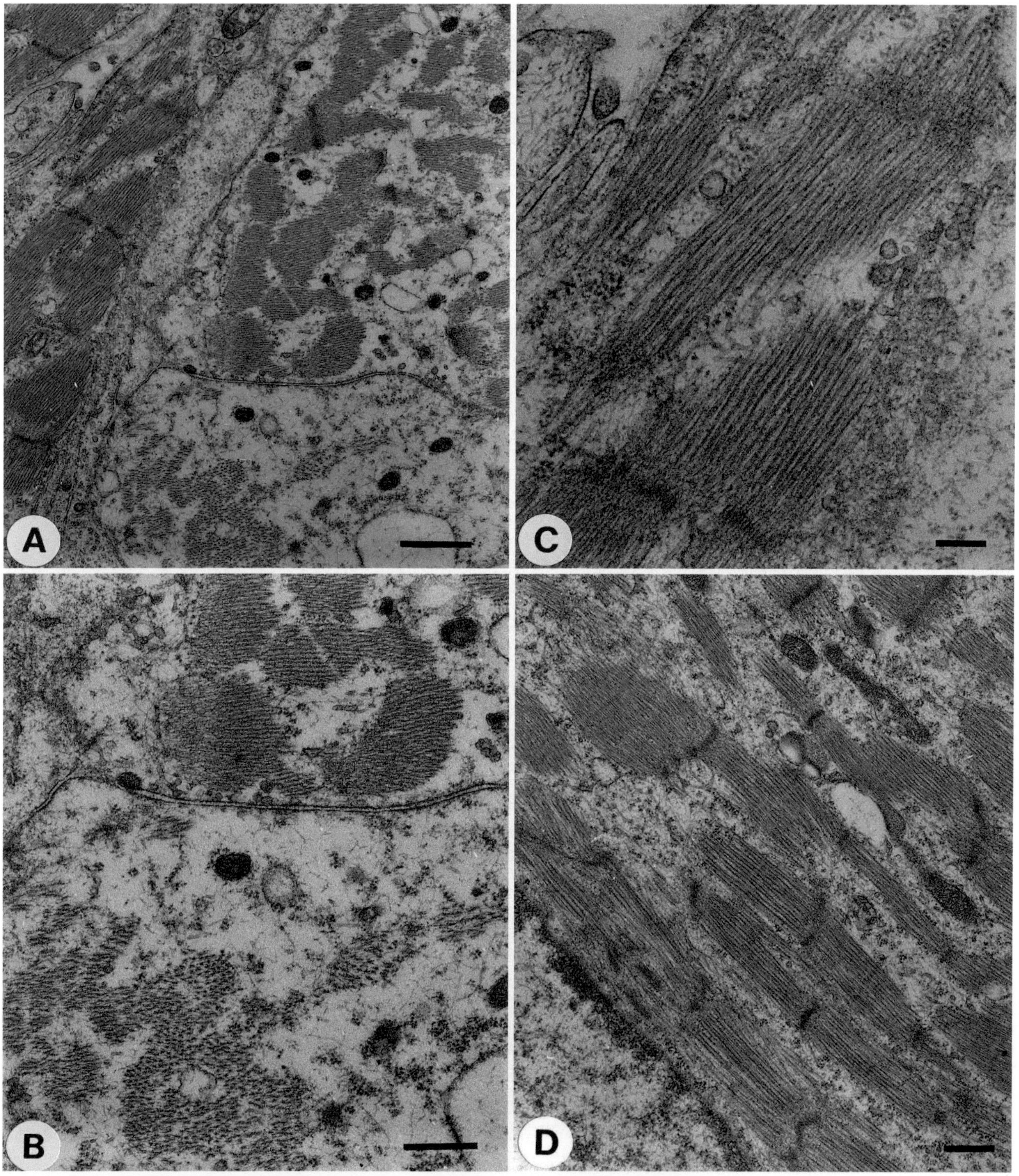

Fig. AI.5. Early development of dermal muscle fibers. Dorsal skin of day 2 rats was used for the experiments. *Bars* **A** 1 μm; **B** 0.5 μm; **C** 0.2 μm; **D** 0.5 μm

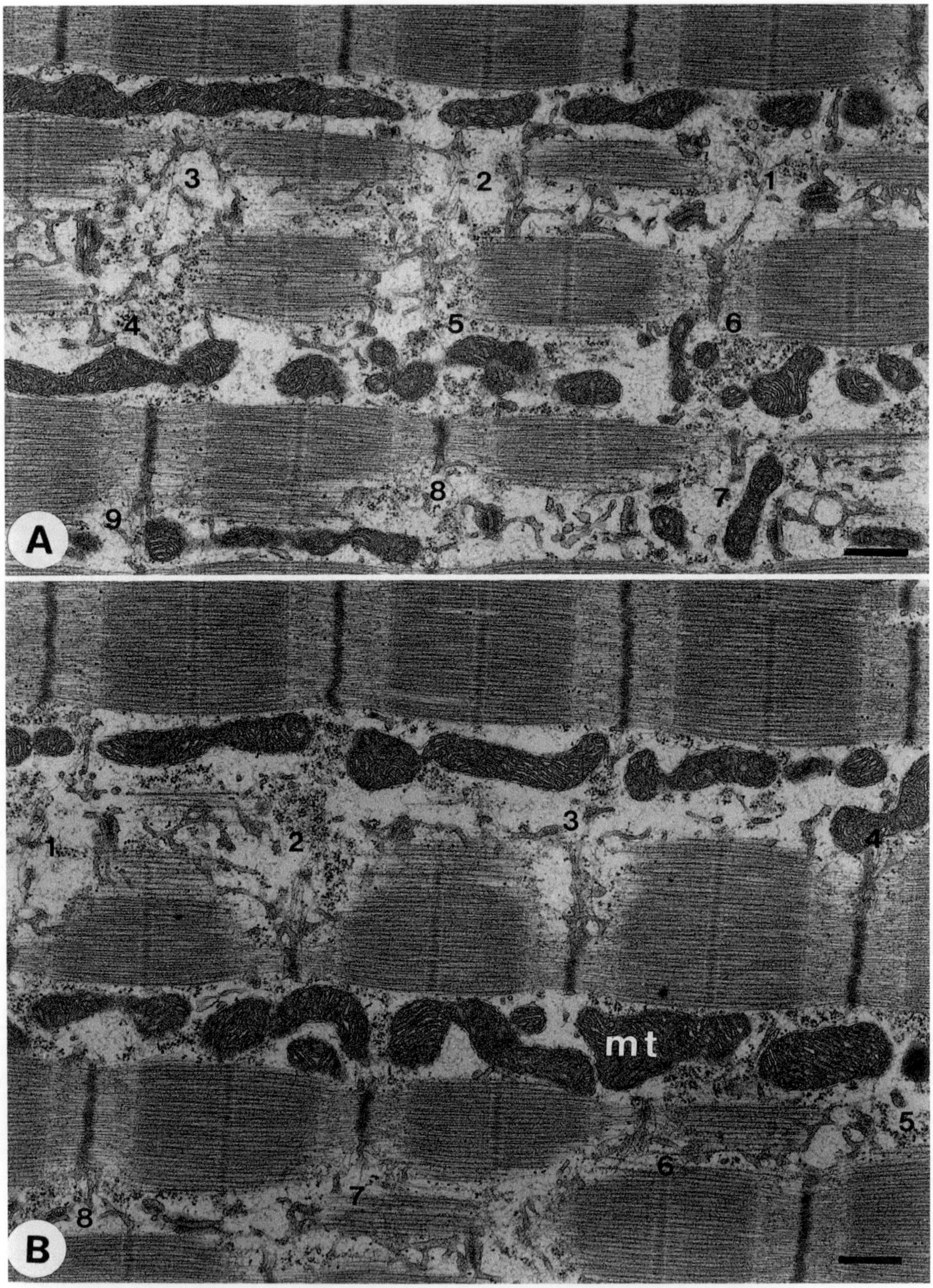

Fig. AI.6. Formation of sarcomeres. Unlike Fig. AI.5, the orientation and the localization of the sarcomere unit are fixed, and it is formed in order as *1* to *3*, *4* to *6*, and *7* to *9* in **A** or *1* to *4*, *5* to *8* in **B**. The formation of the Z band was demonstrated especially well. Dorsal skin of day 14 rats was used. *Bars* 0.5 μm

Dermal Capillaries

Hashimoto has shown the presence of a number of fenestrations in the venules of the perifollicular vascular plexus [10]. In neonatal rats, capillaries are sparsely distributed in the dermis and do not surround the hair follicle. Examples are shown in Fig. AI.7. The hair follicles of fetal and neonatal animals are required to grow under conditions of poor nutritional and oxygen supply. The sensory hair follicles seem to be vascularized in preference to the pelage hair follicles in the facial dermis. Within 1 to 2 weeks of birth, the embryonic nucleated erythrocytes are replaced by intermediate or adult-type enucleated erythrocytes [11] (Fig. AI.7).

Adipocytes

The adipocytes are major components of the subcutaneous tissue. Most hair bulbs are surrounded by adipocytes, although they are usually partitioned by the thin cytoplasm of connective tissue follicle cells (e.g., Fig. AI.8A). The adipocytes differentiate as their lipid droplets grow and occupy their cytoplasm (Fig. AI.8). The lipid droplets are classified into two species, tentatively termed type 1 (smooth) and type 2 (cracked) on the basis of electron microscopic appearance (Figs. AI.8B, AI.9A). Eventually, most of the cytoplasm is filled with lipid, except the periphery (Fig. AI.9A), where autophagy may still occur (Fig. AI.9C). Ultimately, the cell membrane is broken and the peripheral cytoplasm lost (Fig. AI.9B), and only the lipid droplet is left (Fig. AI.9D). Although the regulatory relationship between adipocytes and hair cells is unknown, nutritional support by the adipocytes is expected to be important to hair growth, especially in the early stages before angiogenesis is sufficient.

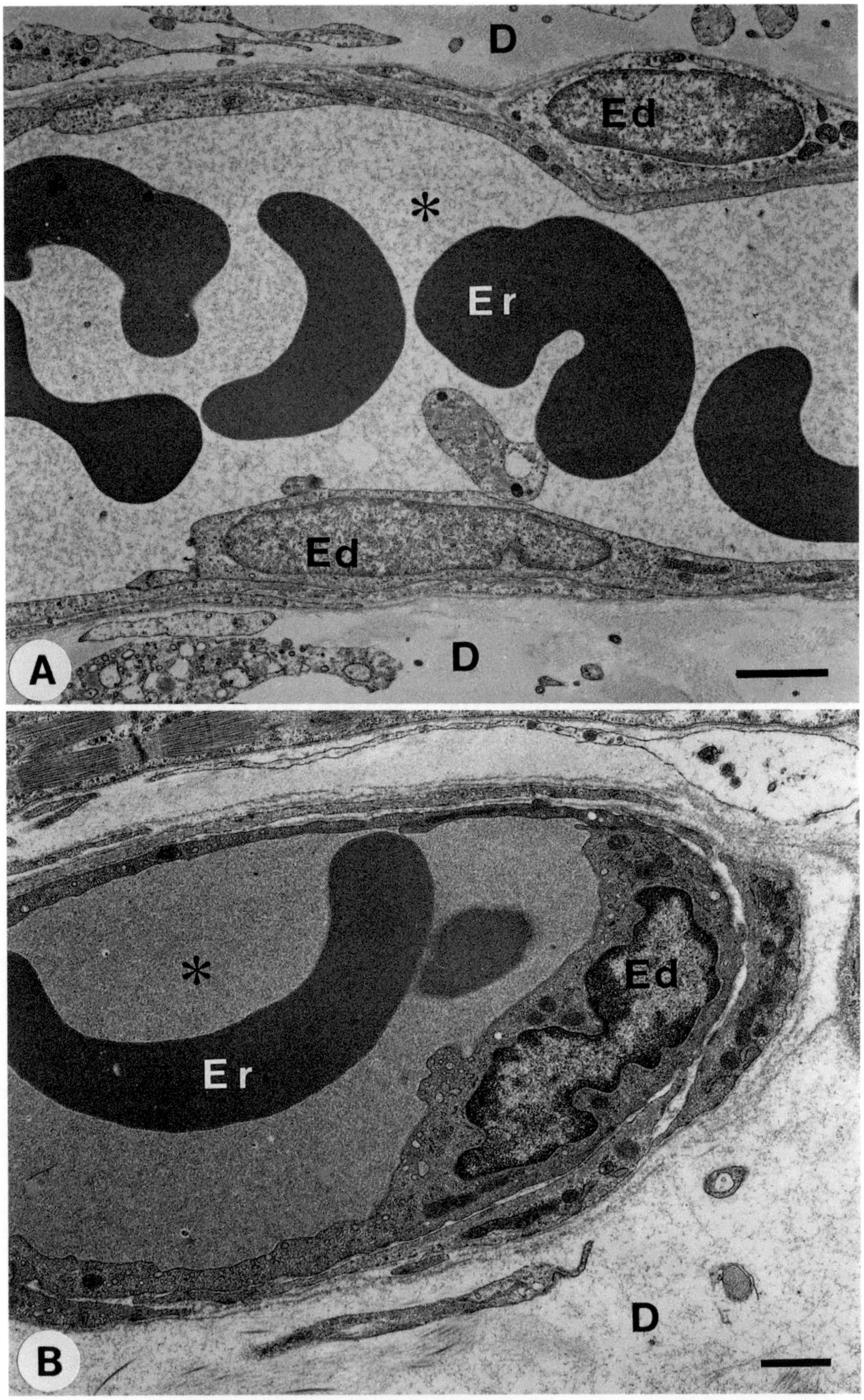

Fig. AI.7. Dermal capillary vessel. *D*, dermal matrix; *Ed*, endothelial cell; *Er*, erythrocyte; *, lumen. The capillaries are gradually distributed in the dermis, although their development is too slow for early hair growth. Later they surround the hair follicle to support the growth. Dorsal skin of a day 7 rat was used in **A** and that of a day 14 rat was used in **B**. *Bars* **A** 2 μm; **B** 1 μm

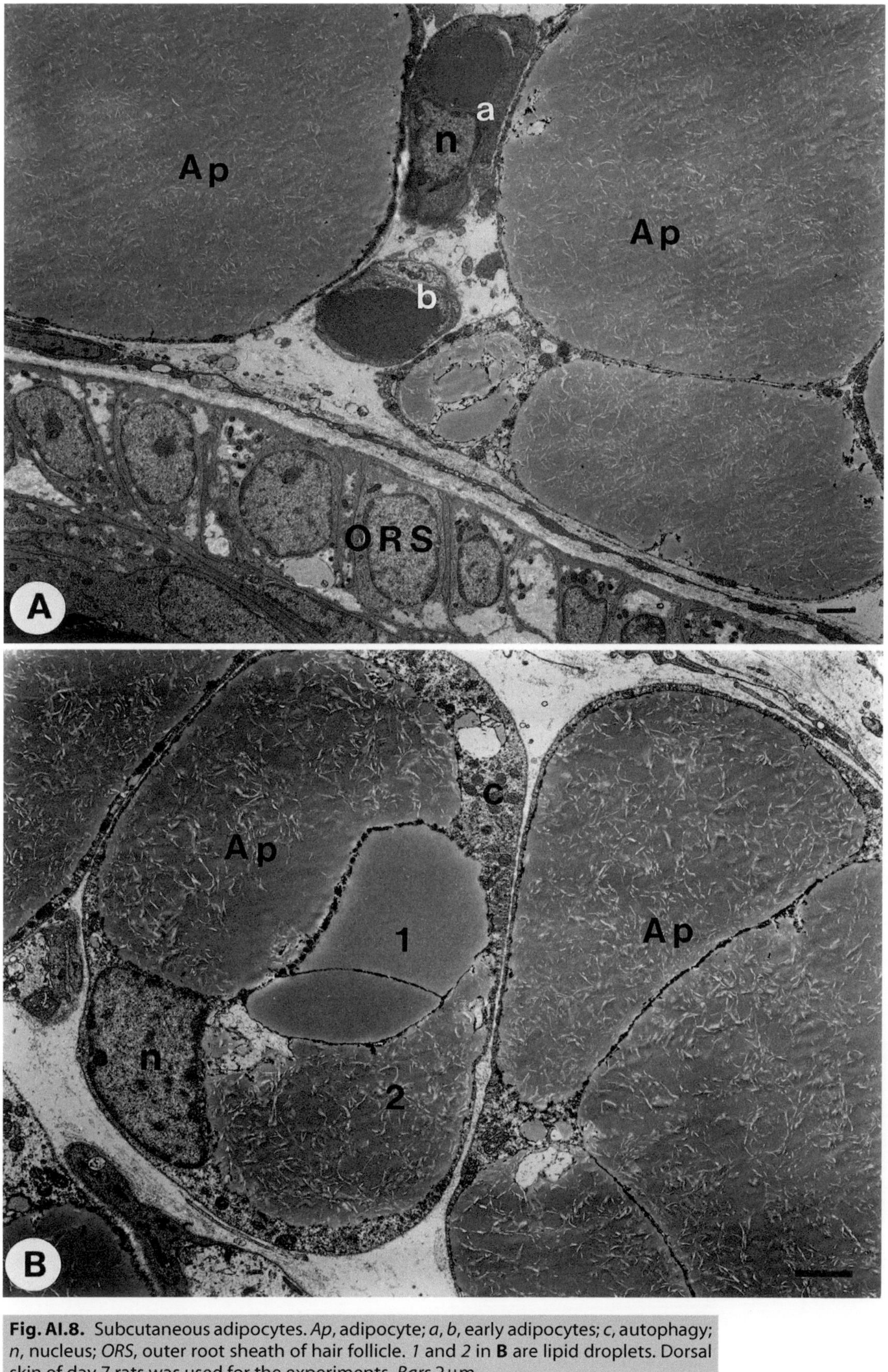

Fig. AI.8. Subcutaneous adipocytes. *Ap*, adipocyte; *a*, *b*, early adipocytes; *c*, autophagy; *n*, nucleus; *ORS*, outer root sheath of hair follicle. *1* and *2* in **B** are lipid droplets. Dorsal skin of day 7 rats was used for the experiments. *Bars* 2 μm

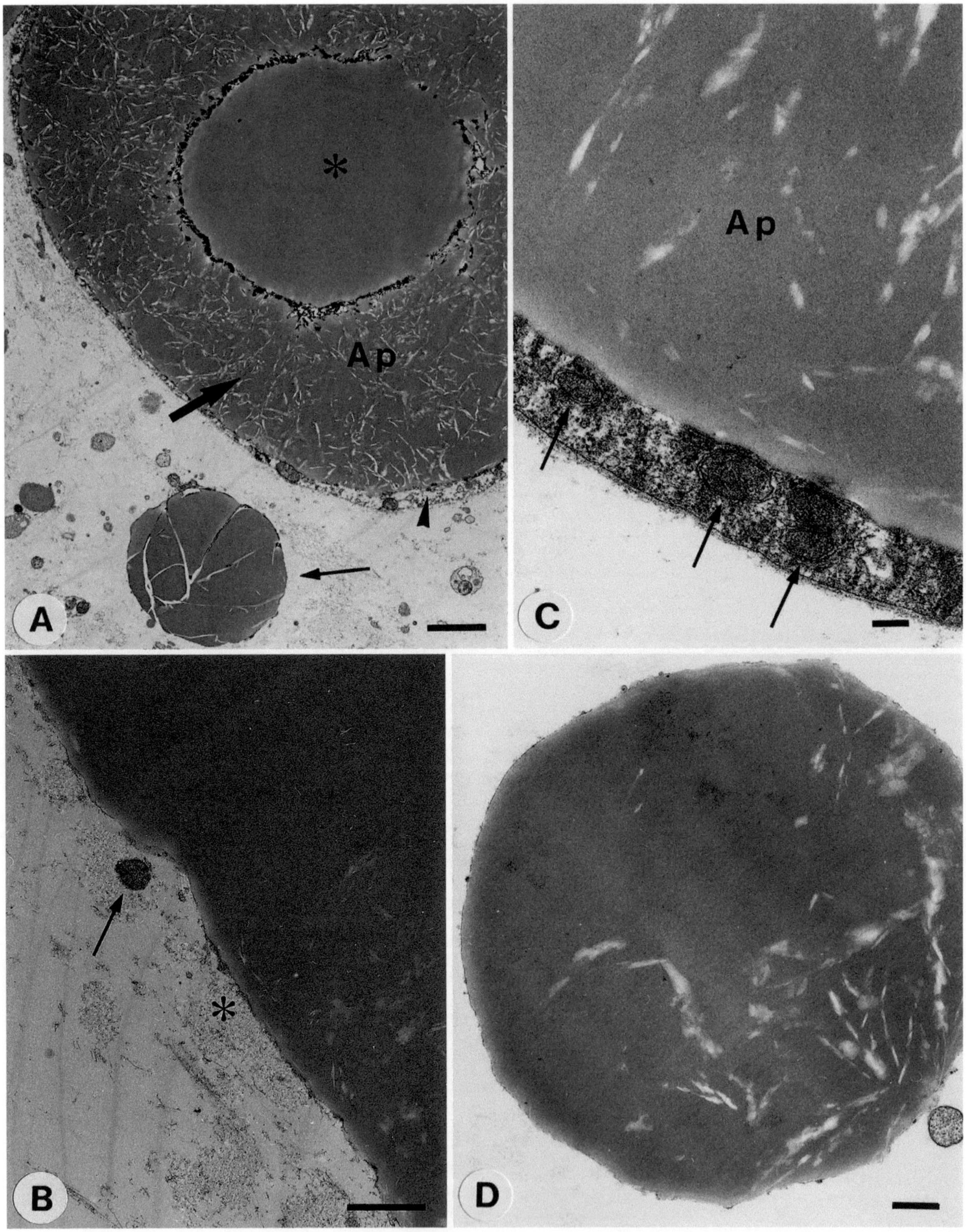

Fig. AI.9.

Neurons and Related Cells

Some follicles of the pelage hairs of rats at day 7 after birth are already in connection with peripheral nerve fibers (Fig. AI.10A). The established adult follicles are often surrounded by peripheral nerves, highlighting the role of hair as not only an insulator but also a sensory organ. It is of great merit for an organism to be able to feel a thing before its direct contact with the skin, because it may be poignant or dangerous. Grooming in particular is an important means of communication in mammals, and grooming or touch by the mother often confers feelings of comfort, relief, and happiness. The angle of the hair, determined via arrector pili smooth muscles under nervous system control, is important to the maintenance of temperature. Innervation is thus essential to the functioning of the hair follicle. Both myelinated and unmyelinated axons innervate the vicinity of hair follicles. A group of these axons and Schwann cells is often encased by the dermal mantle cells (Fig. AI.11A). Several steps in the differentiation of peripheral nerves are shown in Figs. AI.10 and AI.11.

Fig. AI.9. Terminal differentiation and cell death of adipocytes. *Ap*, adipocytes. **A** Terminally differentiated adipocytes. *Thick arrow*, type 2 lipid droplet; *, type 1 lipid droplet; *thin arrow*, free lipid droplet; *arrowhead*, cytoplasm of adipocytes. **B** Death of the adipocytes. *Arrow*, degenerated mitochondria; *, digested cytoplasm of adipocytes. **C** Autophagy in adipocytes. *Arrows*, autophagosome. **D** Free droplet. Dorsal skin of day 7 (**A**) and day 2 (**B**, **C**, **D**) rats was used for the experiments. *Bars* **A** 2 μm; **B** 1 μm; **C** 0.2 μm; **D** 0.5 μm

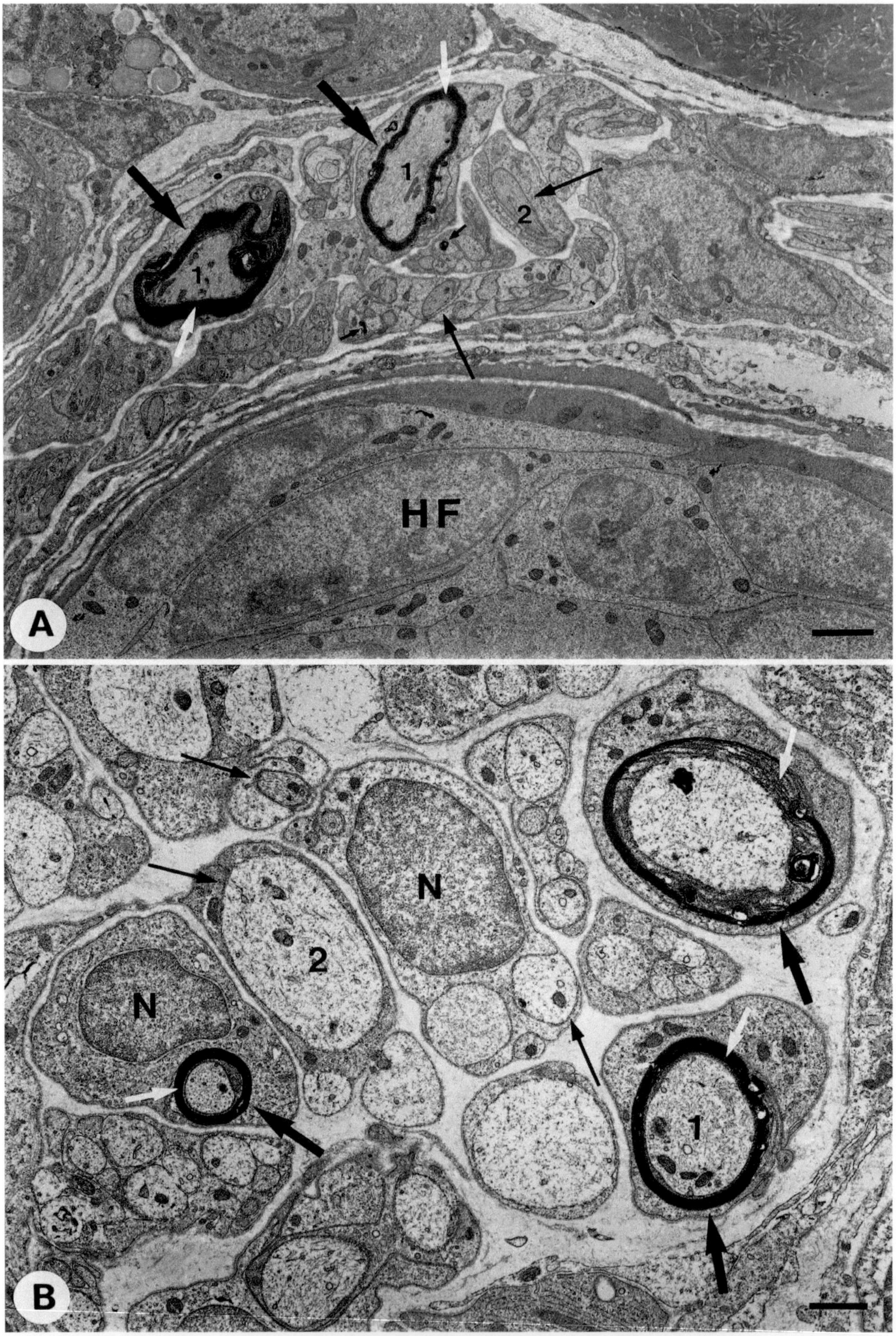

Fig. AI.10. Peripheral nerve system innervated around hair follicle (*HF*). *1*, Myelinated axon; *2*, unmyelinated axon; *thick arrows*, Schwann cell encasing myelinated axon; *thin arrows*, Schwann cell encasing unmyelinated axon. Dorsal skin of day 7 rats was used for the experiments. *Bars* **A** 2 µm; **B** 1 µm

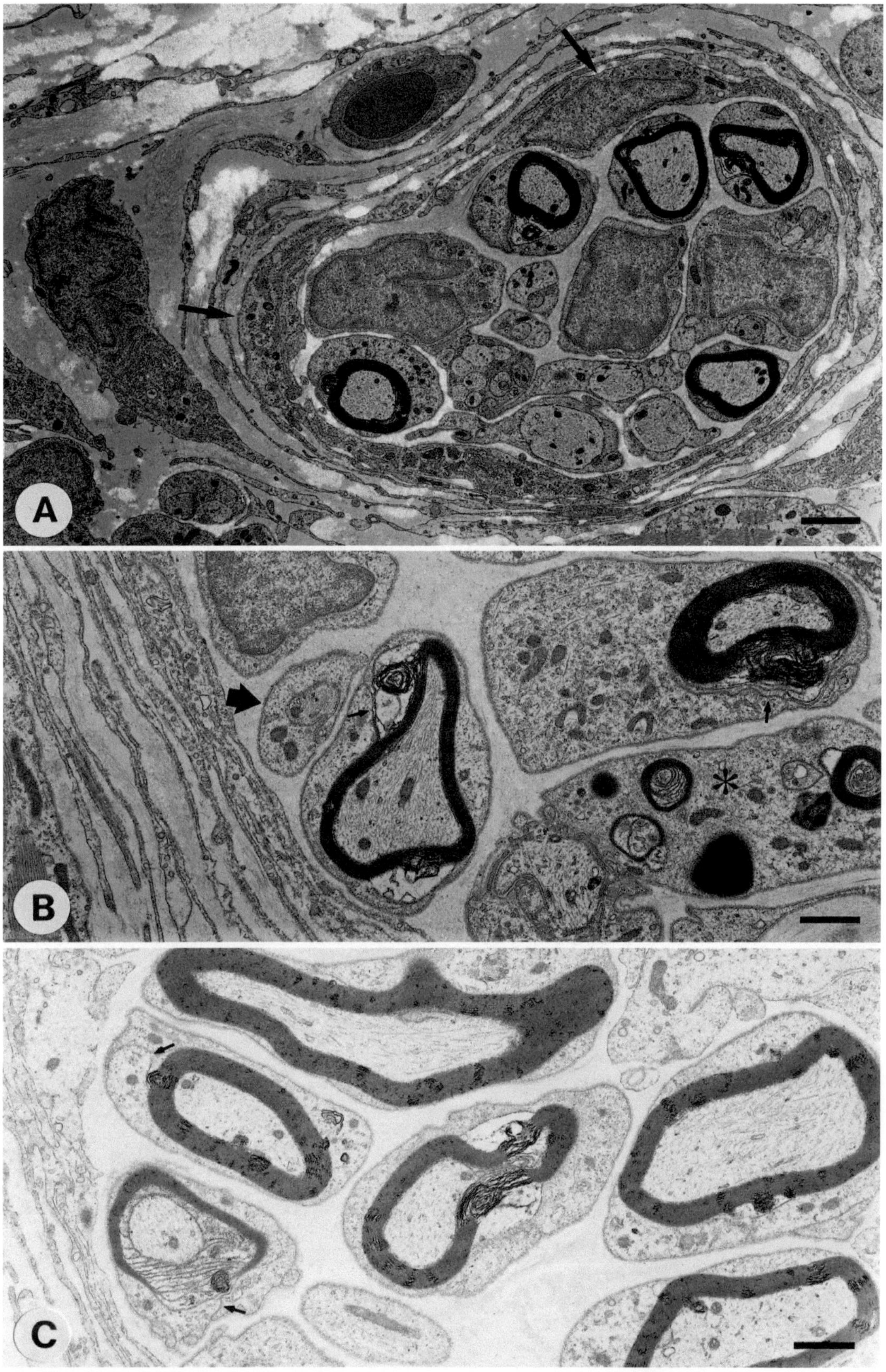

Fig. AI.11. Peripheral nerve system in dermis. **A** *Thin arrow*, dermal mantle cell. A Schwann cell encasing unmyelinated axon with a mesaxon. **B** *Small arrow*, mesaxon in a Schwann cell encasing myelinated axons. *, Schwann cell encasing both myelinated and unmyelinated axons

References

1. Hardy MH (1992) The secret life of the hair follicle. Trends Genet 2:55–61
2. Jamora C, DasGupta R, Kocieniewski P, Fuchs E (2003) Links between signal transduction, transcription and adhesion in epithelial bud development. Nature (Lond) 422:317–322
3. Metcalfe DD, Baram D, Mekori YA (1997) Mast cells. Physiol Rev 77:1033–1079
4. Takano-Ohmuro H, Oishi K, Uchida MK, Watabe M, Morioka K, Kitani S-I, Shikonin (β-alkanine) inhibits protein kinase C activity and histamine release from human basophils. (Manuscript in preparation)
5. Cross PC, Mercer KL (1993) Cell and tissue ultrastructure. A functional perspective. Freeman, New York, p 76
6. Fox RS (2002) Comparative vertebrate anatomy, biology 308, musculature 2, E-2. http://www.lander.edu/rsfox/308muscle2lec.html
7. Imanishi N (2002) Terms in the field of epidermis, muscle-bone system, and respiratory system (in Japanese). Aimic 23:19–26
8. Fawcett DW (1986) A textbook of histology. Saunders, Philadelphia
9. Wolfe SL (1993) Molecular and cellular biology. Wadsworth, Belmont, pp 471–473
10. Hashimoto K (1988) The structure of human hair. Clin Dermatol 6:7–21
11. Morioka K, Minamikawa-Tachino R (1993) Temporal characteristics of the differentiation of embryonic erythroid cells in fetal peripheral blood of the Syrian hamster. Dev Growth Differ 35:569–582

Keratins

Appendix II

Keratins

Although the structural variety of keratinization as revealed by the electron microscope cannot yet be explained on the basis of the three-dimensional assembly of keratin molecules and associated proteins, it is possible to list already known hair keratin species and their localization. The keratin multigene family consists of "soft" and "hard" keratins, the former expressed in various epithelial tissues and the latter restricted to hair, nail, horn, and other hard tissues. Keratins of both types can be divided into type I (acidic) and type II (basic to neutral) members. The association of equimolar amounts of types I and II keratins is obligatory to the formation of intermediate filaments. The keratin dimer is suggested to consist of a coiled coil of type I and type II keratins aligned in register and in parallel fashion, and the tetramer to consist of two dimers in antiparallel fashion, without polarity [1, 2].

The keratin molecule, similar to other proteins forming intermediate filaments, consists of globular amino and carboxyl terminal areas and a rodlike central region (Fig. AII.1). Although definitive evidence has not been obtained, the two ends of the keratin molecule are both thought to be responsible for the tandem adhesion, while the rodlike, alpha helix-rich region is responsible for the lateral association in the forming of the coil. As are other intermediate filaments, the keratin molecules are organized to form a filament of 10 nm diameter (Fig. AII.1). Readers will frequently find the term "tonofilament" in preceding chapters. The tonofilament is the typical keratin-based intermediate filament. In contrast to microfilaments or microtubules, which undergo both synthesis and degradation at the same time to maintain balance, tonofilaments are highly stable once synthesized. Furthermore, the "hard" keratins are known to be stabilized by numerous disulfide linkages in their globular end regions. The stability of tonofilaments contrasts with the dynamic nature of microfilaments or microtubules, although in this regard the drastic assembly of the keratin filaments should not be forgotten (see Chapter 4).

A recent study suggested that the intermediate filaments are not so stable as they had been expected [4].

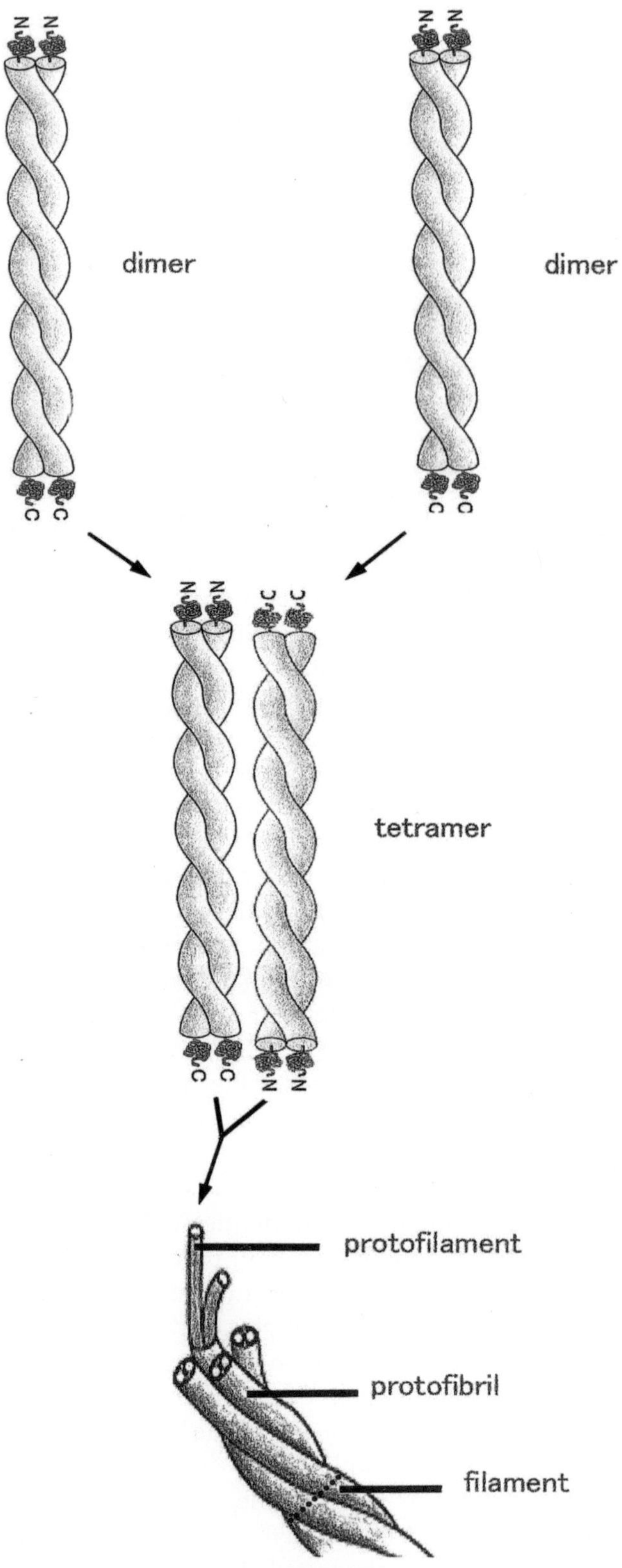

Fig. AII.1. A model of the assembly of keratin molecules. A keratin molecule consists of a central rodlike domain and two globular end domains. The monomer assembles into the *dimer*, in which the genetically conserved rod domains are aligned in parallel and are wound together into a coiled coil. Two dimers then line up side by side in an antiparallel manner (as a *tetramer*) to form a *protofilament* of diameter about 3 nm. The final *filament*, with diameter about 10 nm, is thought to be composed of 8 protofilaments (32 keratin molecules) joined end on end to its neighbors to form the long fiber. In the filament, the rodlike domains interact with one another to form the uniform core, while some parts of the globular domains may project from the filament surface

The catalogue of known hair keratins is shown in Table 1.2. As may be feared, the terminology of keratins is in a critical state. Because K9 and the higher numbers are occupied by type I keratins, newly found type II keratins overflow out of the K-series terminology. Some of these overflows were reluctantly termed—irs,—hf, and so on (see Table 1.2). Powell and Rogers [3] have proposed the addition of 1. for type I keratins and 2. for type II keratins, so that for example K14 becomes K1.14 and K5 becomes K2.5. This system is a good idea, although it is not used in Table 1.2. The variety exhibited by different tissues in the nature of their keratinization should be considered on the basis of the characteristics of the individual keratin molecular species and their combinations, notwithstanding that the means of their higher order construction and molecular assembly have not been fully elucidated. It should also be noted that keratin cables consist of not only keratin-based filaments but also keratin-associated proteins as well as other unknown structural or regulatory components.

Recent advances in molecular biology have promoted research into keratin genes and their expression. For the reader's convenience, I list the hair keratin species described to date and their localization in Table 1.2. The keratin multigene family consists of "soft" and "hard" keratins. Generally, hard keratins contain a number of cystines for the formation of $^{-}$s–s^{-} bonds in their globular end regions. These intermolecular bonds serve to stabilize the hair structure. Both soft and hard groups of keratins are divided into type I (acidic) and type II (basic and neutral) members.

References

1. Coulombe PA, Fuchs E (1990) Elucidating the early stages of keratin filament assembly. J Cell Biol 111: 153–169
2. Steinert PM (1993) Structure, function, and dynamics of keratin intermediate filaments. J Invest Dermatol 100:729–734
3. Powell BC, Rogers GE (1997) The role of keratin proteins and their genes in the growth, structure and properties of hair. In: Jolles P, Zahn H, Hoker H (eds) Formation and structure of human hair. Birkhauser, Basel, pp 59–148
4. Helfand BT, Chang L, Goldman RO (2004) Intermediate filaments are dynamic and motile elements of cellular architecture. J Cell Sci 117:133–141

Subject Index

MIX
Papier aus verantwortungsvollen Quellen
Paper from responsible sources
FSC® C105338

If you have any concerns about our products,
you can contact us on
ProductSafety@springernature.com

In case Publisher is established outside the EU,
the EU authorized representative is:
Springer Nature Customer Service Center GmbH
Europaplatz 3, 69115 Heidelberg, Germany

Printed by Libri Plureos GmbH
in Hamburg, Germany